Guilherme Sa

High-intensity interval exercise causes cardiac remodelling

Guilherme Sa

High-intensity interval exercise causes cardiac remodelling

Beneficial through modulation of local RAS in mice fed a high-fat or high-fructose diet

ScienciaScripts

Cover image: www.ingimage.com

This book is a translation from the original published under ISBN 978-613-9-73169-5.

Publisher:
Sciencia Scripts
is a trademark of
Dodo Books Indian Ocean Ltd. and OmniScriptum S.R.L publishing group

120 High Road, East Finchley, London, N2 9ED, United Kingdom
Str. Armeneasca 28/1, office 1, Chisinau MD-2012, Republic of Moldova, Europe
Printed at: see last page
ISBN: 978-620-7-88559-6

DEDICATORY

To my family, my friends and the whole universe that conspires to make everything go wrong while we fight to make it right.

ACKNOWLEDGEMENTS

To my family, my parents and my friends, who gave me all the support I needed to complete not only my master's degree, but all the achievements of my life.

To my mum who is my strength, who inspires me to fight for her and for myself. None of this could exist if it weren't for her.

The person who came into my life in an unusual way and remains in it to this day, believing in me and accepting me the way I am, without ever asking me to change. Ana Carolina da Silva Paiva, thank you so much!

To eternal friends who are the family we have chosen, brothers and sisters in life, who have always supported and encouraged me, who have always believed in me even though I was nothing and nobody. We will always be together because a friend is not someone who is always physically present, but only a true friend when they need to be.

To the doctor, master and great friend Afonso Aragão, whom I met at university and will carry with me for life. His support and the learning I received was essential for my professional and personal development; without him it wouldn't have been possible to even start this work. I am infinitely grateful for the example of a man that he is, and who has become everything I want to be.

To my great friend Marcel Pimenta for his enormous help throughout this work and the whole of my degree, always by my side at all times.

My friends Thatiany Marinho, Fernanda Ornellas, Vanesa de Souza-Mello, Aline Penna, Michele Soares and Iara Karise, who helped me and were always willing to help whenever I needed it.

To my friend Helder Gonçalves, who proved to be an invaluable friend on my journey and who still hasn't given up on asking for my hand in marriage.

Prof Dr Márcia Barbosa Águila Mandarim-de-Lacerda for her attention and promptness with my work, and for all her advice on the nutritional aspects of the study.

And last but not least, quite the opposite, it's of the utmost importance.

To Profª . Drª . Sandra Barbosa Silva for her trust, support, understanding, help and patience with my work, as well as enabling me to develop it with all the technical support and subsidies inherent in the study.

Fate is inexorable.

Uthred of Bebbanburg

SUMMARY

HIIT (high-intensity interval training) has the potential to reduce cardiometabolic risk factors, but its effects on cardiac remodelling and the local renin-angiotensin system (RAS) in mice fed a high-fat or high-fructose diet have yet to be clarified. Sixty male C57BL/6 mice (12 weeks old) were randomly divided into three groups according to the diet offered (control, high-fat diet (HF) or high-fructose diet (HFRU)) and were monitored for 8 weeks before the HIIT protocol. The animals were then subdivided into six groups (n = 10 each), three groups started a 12-week HIIT protocol, while the other three remained sedentary. HIIT reduced body mass and systolic blood pressure, while increasing insulin sensitivity after ingesting the HF and HFRU diets. In addition, HIIT reduced the left ventricular hypertrophy caused by both diets. In particular, HIIT positively modulated the key factors required in the left ventricular renin-angiotensin system: reduced protein expression of renin, ACE and AT2 receptor in both trained groups and reduced protein expression of the AT1R receptor only in the trained HF group. HIIT played a beneficial role in modulating the ACE2 / Ang (1-7) / rMas receptor axis. ACE2 gene expression was increased in the HIIT groups, accompanied by greater Mas receptor gene expression after the HIIT protocol. The present study shows the effectiveness of HIIT sessions in producing significant improvements in insulin resistance and attenuating left ventricular hypertrophy, although hypertension was controlled only in the HF diet group submitted to the HIIT protocol. The local RAS system in the heart mediates these findings and the MAS receptor seems to play a crucial role when it comes to improving structural and functional cardiac remodelling due to HIIT.

Key words: High intensity exercise. HIIT. Cardiac hypertrophy. RAS

SUMMARY

INTRODUCTION

The prevalence of obesity has increased worldwide and is associated with our current lifestyle, which promotes sedentary habits and poor eating habits such as excessive fructose intake, mainly in industrialised drinks and foods (Rippe and Angelopoulos, 2016a; Alwahsh and Gebhardt, 2017), or the Western diet, consisting of excessive carbohydrates and fat. These conditions are contributing to the currently increased incidence of cardiovascular disease (CVD) (Rippe and Angelopoulos, 2016b; Sacks et al., 2017).

Physical inactivity is harmful to cardiovascular function and has been associated with an increased risk of chronic diseases such as type 2 diabetes and hypertension (Lakka et al., 2003). Furthermore, nowadays, lack of time is one of the main barriers to adherence. In this context, high-intensity interval training (HIIT) presents itself as a mode of training that requires little time and can be described as "brief intervals of vigorous physical activity interspersed with periods of low activity or rest", which induces a strong acute physiological response (Gibala et al., 2012). In addition, HIIT has been shown to reduce cardiometabolic risk factors in the same way as traditional moderate intensity continuous training (MICT) (Burgomaster et al., 2008; Gibala et al., 2012).

In relation to diets, there is currently greater interest in the potential role of added sugars, especially fructose, as a contributing factor to CVD. When consumed in high concentrations, fructose can promote metabolic changes such as hyperuricaemia, inflammation (Kanuri et al., 2011) and hypertension (Kanuri et al., 2011).

It is important to note that excessive dietary fat intake leads to increased lipid deposition in adipose and non-adipose tissues. In addition, dietary fat induces fatty acid oxidation and consequently increases lipid peroxidation products, promoting a chain of events that leads to the development of CVD (Liang et al., 2014) such as left ventricular hypertrophy (LVH) and diastolic dysfunction (Sverdlov et al., 2016). Similarly, a diet with high fructose concentrations negatively affects cardiac structure and function (Huang et al., 2016).

On the other hand, physical exercise is strongly associated with reduced risk of chronic disease and has been widely used to combat risk factors for obesity, hypertension and CVD (Bassuk and Manson, 2005; Bidwell et al., 2014). In this context, recent evidence indicates that HIIT provides a stronger stimulus than MICT to bring about myocardial improvements (Cassidy et al., 2017).

However, while the relationship between aerobic exercise and blood pressure and cardiovascular health has been widely explored (Fisher et al., 2015; Pimenta et al., 2015), little is known about the direct effects of HIIT on the molecular components of the renin-angiotensin system (RAS), especially in models of obesity induced by high-fat or high-fructose diets.

It is known that RAS has broad cardiovascular regulatory effects and its abnormality participates in the generation and development of hypertension (Giles, 2007). Thus, hyperactivation of the RAS contributes to structural and functional myocardial fibrosis and cardiac hypertrophy, and the local RAS in the heart may play a decisive role (Cassidy et al., 2017). RAS hyperactivity also causes volume overload and peripheral vasoconstriction, leading to increased LV diastolic pressures, LVH and changes in cardiac geometry (Chinnaiyan et al., 2005; Sciarretta et al., 2009). We now know that the classic angiotensin-converting enzyme (ACE) / angiotensin II (Ang II) / angiotensin 1 (AT1) receptor axis is not the only signalling pathway involved in RAS activation, but other pathways such as the ACE 2 / Ang (1-7) / Mas receptor (rMAS) axis play a key role opposing the effects of Ang II on the cardiovascular and renal systems (De Mello, 2017).

Considering that animals fed high-fat or high-fructose diets are well-established rat models used to study cardiac hypertrophy / cardiomyopathy (Raher et al., 2008; Bouchard-Thomassin et al., 2011), with insulin resistance, SAH and RAS hyperactivity, and the fundamental role that RAS imbalance plays in adverse cardiovascular remodelling, we hypothesised that HIIT attenuates cardiovascular remodelling, 2011), with insulin resistance, SAH and RAS hyperactivity and the fundamental role that RAS imbalance plays in adverse cardiovascular remodelling, we hypothesised that HIIT attenuates adverse cardiovascular remodelling and counteracts the metabolic alterations resulting from excessive fat or fructose intake in mice by modulating local RAS. To this end, we assessed protein and gene expression of the RAS system in the LV, as well as LV wall morphometry; lipid profile, systolic blood pressure and insulin sensitivity were also assessed to determine how HIIT impacted cardiometabolic parameters after fructose or high-fat diet ingestion.

CHAPTER 1

OBJECTIVES

1.1 General

A To qualitatively and quantitatively analyse the beneficial mechanisms of HIIT on the cardiovascular system of C57BL/6 mice subjected to a high-fructose, high-fat diet.

1.2 Specific

- To assess basal metabolism and biometric parameters before and after the training protocol;
- Analysing cardiovascular remodelling using stereological tools;
- To evaluate the effects of high-intensity training on carbohydrate metabolism, lipids and blood pressure;
- To determine the gene and tissue expression of heart-specific proteins in order to evaluate the local activation of the renin-angiotensin-aldosterone system.

CHAPTER 2

LITERATURE REVIEW

2.1 Heart disease

Heart disease is the term that covers all diseases that affect the heart, including myocardial disease, congenital heart disease, heart infections, valvular heart disease, ischaemic heart disease and hypertensive heart disease (Kalla et al., 2016).

Epidemiologically, according to the WHO, cardiovascular diseases are responsible for 30% of mortality, with an individual having the possibility of developing heart disease in an estimated 15% over a period of 10 years (SIGN, 2007). There is a 90 per cent chance of developing an acute myocardial infarction (AMI) if smoking habits are not changed. And yet, the disabilities caused by cardiovascular diseases are likely to increase from 85 million patients to 150 million patients by 2020 (Ko et al., 2006; SIGN, 2007).

In Brazil, it is estimated that 345,111 people died from some type of cardiovascular disease, with an increase in mortality of 82.73% in the period 2004-2015 (SBC, 2015). In this context, there is a predominance of deaths from cardiac ischaemia (especially AMI), which reached 54.6 people per 100,000 inhabitants in 2013. This year also saw a peak of 23.9 deaths per 100,000 inhabitants caused by hypertensive diseases, especially systemic arterial hypertension (SAH) (SBC, 2015). In 2011, Mansur and Favarato (Mansur and Favarato, 2012) identified hypertension as the main cause of death among men and women in Brazil, accounting for 20% of deaths in individuals over 30 years of age.

In line with this data, cardiovascular diseases in children are becoming increasingly common, affecting between 8 and 10 children in every 1,000 born, according to Junior et al. 2002 (Pinto JR et al., 2004). This data makes it clear that there is a need for a policy of increasing intervention in this type of cardiopathological manifestation.

Within these pathologies we have a very important aggravating factor that is growing more and more, namely SAH, which is one of the most common causes of AMI and stroke (MS, 2006). A survey carried out in 2015 in the United States of America found that SAH was present in 69% of patients who developed an AMI, 77% who developed a stroke, 75% with heart failure (HF), and 60% with peripheral arterial disease (PAD) (MS, 2006).

In Brazil it is estimated that SAH affects around 36 million adults (32.5%) and is one of the main contributors to death from cardiovascular disease (SBC, 2016). From 2004 to 2013 there were

411,529 deaths from hypertensive diseases or SAH, equivalent to 3.74% of the total number of deaths in Brazil in this period (SBC, 2015). According to statistical data, death from heart disease increases progressively as blood pressure (BP) rises above 115/75 mmHg (SBC, 2010). Estimates in Brazilian cities over the last 20 years have indicated a prevalence of SAH of over 30 per cent.

Among the prevalent risk factors in the aetiology of SAH are overweight and obesity where, even among physically active individuals, an increase of 2.4kg/m^2 in body mass index increases the risk of developing SAH. Salt intake and a sedentary lifestyle are also powerful factors in the prevalence of SAH (SBC, 2010).

2.2 Physical exercise

Nowadays, exercise is very much in vogue as a way of improving health, relaxing, having fun, escaping reality, socialising, among many other possibilities (Anderson and Shivakumar, 2013). Within this myriad, the type of exercise we're focusing on is that which is practised in favour of health.

The term exercise alone does not reveal the real intention of practising it in the search for better health. What must be understood is that in order to characterise the type of exercise we will be doing in order to achieve the beneficial results we hope for in terms of health, we need to know how it is used. By definition, "physical activity" is a generic expression that encompasses any type of movement carried out by the locomotor system that involves greater energy expenditure than rest. "Physical exercise", on the other hand, is any movement of the locomotor system that is carried out in a planned way with the aim of maintaining or increasing health, thus optimising physical fitness (Mochcovitch et al., 2016). We will therefore use the term "physical exercise" in this study.

Similarly, physical exercise has been shown to be a powerful non-pharmacological treatment tool for MS (Boudet et al., 2016), (Lee et al., 2016a). Exercises such as walking, running, cycling, resistance activities, calisthenics, among many other forms of exercise have enormous functionalities and health benefits.

Vehí and colleagues (2016) (Vehi et al., 2016) found that a Nordic walking protocol (a type of walking using a stick in each hand to propel the body) performed twice a week for a period of one year improved the patients' perception of effort, as well as reducing their average body mass from 92.22 ±15.88 to 90.34+17.77 kg, and all the markers relating to the diagnosis of MS: glucose and glycated haemoglobin, LDL and HDL cholesterol, systolic and diastolic blood pressure and triglycerides.

Corroborating these results, Lee et al (2016) (Lee et al., 2016a) identified a greater reduction in the parameters that define MetS in individuals who practised physical exercise once a week or more than those who did not practise any type of physical activity. They also identified that the type and frequency of physical exercise adopted directly influences the results on MetS.

Pierard and colleagues (2016) (Pierard et al., 2016) subjected rodents to treadmill training in which they reached a maximum time of 60 minutes at the end of the training and identified a reduction in MC, glycaemia, and adiponectin receptors 1 and 2.

In another study, Mitranun et al. (2014) (Mitranun et al., 2014) compared groups of adults with type 2 diabetes who underwent continuous training and a second group who underwent high intensity interval training (HIIT). Both used the treadmill as a form of exercise. At the end of the training protocol, a reduction was identified in all MetS components, levels of glycaemia and cholesterol (LDL and HDL), triglycerides, glycated haemoglobin, a reduction in insulin resistance, MC and body mass index (BMI) and levels of systolic and diastolic blood pressure for both trained groups; at the end of the study, there was a balance of beneficial responses from both groups to the training.

In view of the above, we realise that more than one type of physical exercise is an option for the non-pharmacological treatment of MS, each of which is effective in its own way.

2.3 *High-intensity interval training (*HIIT)

Exercise plays a central role in the control and treatment of common metabolic diseases, but modern society presents many barriers to exercise. In adults, the recommended amount of exercise to modify health risks is equivalent to 150 min/week of moderate-intensity continuous exercise (MICT) or 75 min/week of vigorous exercise (Garber et al., 2011). The biggest barrier to regular physical activity is lack of time, which questions the practicality of HIIT in the total adult population, considering their routine commitments (Samir et al., 2011). HIIT can be described as "brief intervals of vigorous activity interspersed with periods of low activity or rest", which induces a strong acute physiological response and has been widely studied over the last decade (Gibala et al., 2012).

Various HIIT protocols have been adopted in the literature, but most interventions use high-intensity intervals of 1 to 4 minutes. The aim of HIIT is to accumulate activity at an intensity that the participant is unable to sustain for prolonged periods (i.e. 80-95% of maximum oxygen consumption (peak V:O2) or > 90% of maximum heart rate (HRmax), whereby the recovery time must be sufficient to allow the subsequent interval to be completed at the desired intensity (Nicolo and Girardi, 2016). The total duration of a HIIT session tends to be < 20 min, which is comparable to the recommendations of exercise moderation organisations in terms of duration. There is also a subcategory of HIIT involving intervals of 10-30 seconds and intensities that often exceed 100% V:O2peak, i.e. "all-out" exercise with a workload above maximum aerobic capacity. This type of exercise is called sprint interval training and has not been substantially tested in clinical populations (Burgomaster et al., 2005).

Most published studies using HIIT, particularly in clinical populations, have used exercise modalities involving cycling, walking and running, mostly performed on stationary cycles and treadmills. However, other equipment such as cross-trainers/ellipiticals are reasonable options for some (Fex et al., 2015). Of course, there is clear variation across the literature and it has yet to be determined whether there is an optimal HIIT protocol for treating metabolic changes.

2.3.1 The role of HIIT on the cardiovascular system

Cardiovascular complications are the leading cause of mortality in those with common metabolic diseases (Rafiq et al., 2009). Designing the HIIT interval to include rest periods allows patients to accumulate time at higher exercise intensities, thus challenging the cardiovascular system. Limited evidence indicates that HIIT provides a stronger stimulus than MICT to induce myocardial improvements. Alongside the beneficial impact of HIIT on vascular and cardiorespiratory fitness, this suggests that the cardiovascular benefit of HIIT outweighs the metabolic benefit (Nicolo and Girardi, 2016).

With regard to the cardiac adaptations acquired through the practice of HIIT, most of the molecular mechanisms are obtained from studies in experimental rodent models, due to the difficulty of obtaining human myocardial tissue. Rodent hearts have similarities to human hearts and mimic humans in their cardiac response to physical exercise (Hasenfuss, 1998; Hollekim-Strand et al., 2014). The db/db mouse model provides a good representation of the human heart in diabetic patients. After 13 weeks of HIIT, contractility and Ca2+ availability were restored to normal levels (Stolen et al., 2009). These adaptations occurred despite no improvement in glucose or insulin concentrations, demonstrating the direct impact of HIIT on the myocardium. Similar adaptations were observed in heart failure models and healthy rodents (Wisloff et al., 2002; Kemi et al., 2005), with greater changes occurring after high-intensity exercise (85-90% of maximal oxygen consumption [V:O2max]) compared to moderate-intensity exercise (65-70% V:O2max) (Kemi et al., 2005).

Exercise also activates phosphoinositol-3 kinase/Akt, a target of rapamycin signal transduction (mTOR) that leads to increased ribosomal biogenesis and protein synthesis, thus inducing physiological hypertrophy HIIT (85-90% V:O2max) versus moderate (65-70% V:O2max) MICT (Wisloff et al., 2002; Kemi et al., 2005).

The activated pathways induced by exercise in disease models may differ (Wisloff et al., 2002), but healthy human and rodent models indicate that exercise stimulates important mechanisms of transcriptional and translational regulation that lead to structural remodelling of cardiac tissue and thus to improved strength of cardiac contractions (Stolen et al., 2009).

In relation to cardiac structure, it is postulated that adults with common metabolic diseases show concentric remodelling of the left ventricle, which represents a reduction in end-diastolic volume and is also known as pathological hypertrophy (Zile et al., 2011). This reduction in end-diastolic volume occurs in response to stress signals and is a reflection of an accumulation of collagen in the myocardium (Frey et al., 2004). HIIT, on the other hand, has been shown to induce physiological hypertrophy (Cassidy et al., 2016), increasing left ventricular wall mass and end-diastolic volume through a physiological response to growth signals (Frey et al., 2004). The number of studies investigating cardiac structure after HIIT is small; a recent study demonstrated an 8 ml increase in end-diastolic volume after 12 weeks of HIIT in patients with type 2 diabetes (Cassidy et al., 2016). In addition, the energy expenditure of HIIT has been shown to be higher than the energy expenditure of MICT, and added to this factor it has caused structural cardiac remodelling in those with hypertension (Molmen-Hansen et al., 2012).

Another important cardiovascular aspect of HIIT is its role in controlling hypertension. Thus, a recent study reported that twelve weeks of HIIT induced an improvement in systolic blood pressure in adults with type 2 diabetes (Hollekim-Strand et al., 2014; Cassidy et al., 2016), hypertension (Molmen-Hansen et al., 2012) and heart failure. Twelve weeks of HIIT in hypertensive patients improved initial events in systole, which correlate with contractility and are independent of load (Molmen-Hansen et al., 2012). These improvements are the same as those observed with commonly used pharmacological treatment, such as ACE inhibitors or beta-blockers (Wisloff et al., 2007).

Figure 1 - Cardiometabolic effects of HIIT. The figure shows the muscular and cardiovascular impact of HIIT on common metabolic diseases

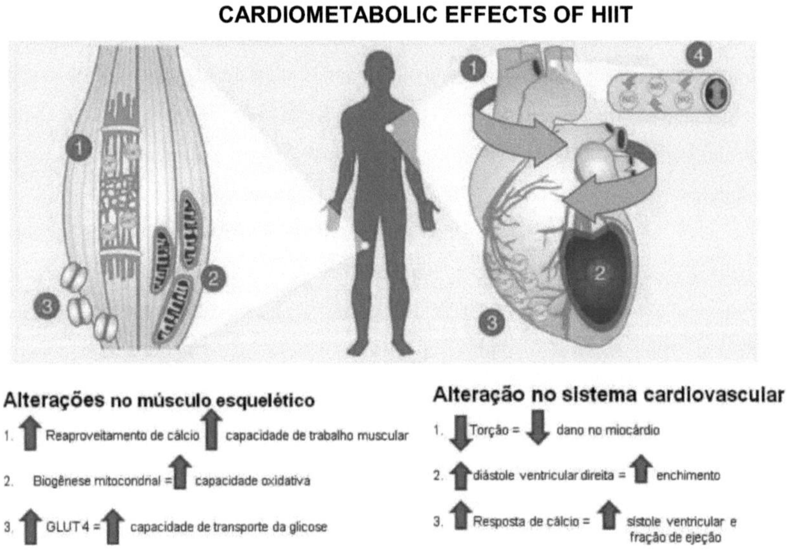

CARDIOMETABOLIC EFFECTS OF HIIT

Alterações no músculo esquelético

1. ↑ Reaproveitamento de cálcio ↑ capacidade de trabalho muscular
2. Biogênese mitocondrial = ↑ capacidade oxidativa
3. ↑ GLUT4 = ↑ capacidade de transporte da glicose

Alteração no sistema cardiovascular

1. ↓ Torção = ↓ dano no miocárdio
2. ↑ diástole ventricular direita = ↑ enchimento
3. ↑ Resposta de cálcio = ↑ sístole ventricular e fração de ejeção

Note: Adapted from Cassidy et al.
Source: (Cassidy et al., 2017).

2.3.2 Cardiac hypertrophy and high-intensity interval training

Cardiac hypertrophy is characterised as a response of the heart to an increase in the demand for functional work, due to an intervening agent causing chronic haemodynamic overload. Many factors are responsible for causing this hypertrophy, but the close relationship with the haemodynamic load leads to pressure deformations and increased tension on the heart wall, leading to an increase in myocardial volume (Franchini, 2001).

As myocardial cells are incapable of dividing in adulthood, their hypertrophy ends up occurring due to pressure adaptations, which leads to a

an increase in myocardial mass (Anversa et al., 1986), resulting in thickening of the wall of the heart chambers, which is more pronounced in the wall of the left ventricle (LV).

It should be pointed out that there are different types of cardiac hypertrophy, thus presenting different structural characteristics where pressure hypertrophy, volume overload and response to physical training are totally different in their functional factors, however, they present the increased haemodynamic load on the heart as a common pathogenic factor (Levy et al., 1990).

When cardiac hypertrophy is related to pathological conditions such as hypertension, it becomes much more harmful to human beings, becoming one of the main risks of morbidity and mortality and can lead to sudden death, systolic and diastolic ventricular dysfunction, ventricular arrhythmia and myocardial ischaemia. As we can see, this hypertrophy in itself is already a powerful risk factor (Mill and Vassallo, 2001). It should be emphasised that it is not only SAH that will be a negative ally for cardiac hypertrophy, but also valvular heart disease, genetic diseases, intracavitary communications, among others.

As we can see, cardiac hypertrophy is one of the great evils of our time, but not all cardiac hypertrophy is bad. There are important differences between pathological hypertrophy and that caused by physical exercise, the latter being known as physiological hypertrophy (Mihl et al., 2008).

Within pathological cardiac hypertrophy there are two different types of hypertrophy with different pathophysiological (harmful) responses: concentric myocardial hypertrophy and eccentric myocardial hypertrophy (Mihl et al., 2008).

According to Katz and Rolett (2016) (Katz and Rolett, 2016) concentric myocardial hypertrophy consists of the synthesis of new sarcomeres arranged in parallel to the existing sarcomeres, thus increasing the force of contraction of the cardiomyocytes. This leads to a thickening of the myocardial wall beyond the limit of the myocardium, with the LV being the most affected cardiac structure. As this pathology progresses, there is a tendency to reduce the thickening of this myocardium, which brings with it a reduction in the heart's contractility, leading to dilated cardiomyopathy. This hypertrophy develops when the LV is overloaded by systemic arterial pressure, for example (Nadruz, 2015).

In eccentric myocardial hypertrophy, the new sarcomeres are arranged in series in relation to the existing ones, which leads to an elongation of the cardiomyocytes, generating an increase in the volume of the LV cavity and a reduction and inefficiency of myocardial contractility (Matos-Souza et al., 2008; Barsukov et al., 2015). It should be borne in mind that the stimuli causing this condition originate from volume overload such as mitral regurgitation.

Physiological hypertrophy occurs as a response to human growth, during pregnancy and through physical exercise (Matos-Souza et al., 2008). The adaptations generated will have a beneficial effect by optimising the functioning of the cardiovascular system, since they will occur through haemodynamic demands arising from physical exercise, among other things (Mangold et al., 2013). These adaptations respond with stretching and better tension in the heart cells, which generates an increase in the myofibrils of the cardiomyocytes (Mihl et al., 2008).

People who do similar hard physical exercise, such as cycling and running, have a greater left ventricular volume with an increase in heart rate, stroke volume, cardiac output and blood pressure, generating an overload in this LV. The heart's response to this demand is to increase the chamber and thickness of the LV wall, with new sarcomeres arranged in series in relation to the existing ones. This change appears beneficial as it has no deleterious effects and does not cause pathological complications (Mihl et al., 2008; Scharf et al., 2010).

In resistance exercisers such as bodybuilders, cardiological adaptations show an increase in the thickness of the ventricular wall, but never beyond physiological limits, with the sarcomeres arranged in parallel to the existing ones, and there is a slight increase in the ventricular chamber. These adaptations are due to the stimulus generated by intermittent periods of increased BP, HR, stroke volume and cardiac output (Mihl et al., 2008).

It should be noted that even though the arrangement of sarcomeres appears similar in pathological and physiological hypertrophy, there are crucial differences in the arrangement of these sarcomeres, since in pathological cardiac hypertrophy they remain agglutinated (thus impairing myocardial function) compared to the sarcomeres in physiological hypertrophy, whether they are arranged in parallel or in series (Matos-Souza et al., 2008).

Whalley et al (2004) (Whalley et al., 2004) in a study carried out in Auckland, New Zealand, used groups of young and elderly men, dividing them into trained and untrained, using as a criterion for trained individuals who regularly practised physical exercise for more than two years and the untrained (NT), the opposite of this and free of heart disease. The elderly group (60 - 80 years old) had 18 NT men and 18 trained men, and the young group (20 - 30 years old) had 12 NT men and 10 trained men. Despite their age, the authors considered both the elderly and the young in the same NT and trained group and did not identify disparate results due to age group variation. The trained individuals showed better morphometric and exercise tolerance parameters than the untrained ones;

they also identified an increase in LV end-diastolic and end-systolic diameter in the trained individuals, as well as an increase in LV mass, but no change in LV wall thickening.

Radovits and colleagues (2013) (Radovits et al., 2013) used young wister rats as an experimental model of swimming training, subjecting the animals to 12 weeks of training with 200 minutes per training session, 5 days a week. The groups were divided into trained and NT, and trained/trained and untrained control groups. It was found that swimming training, when compared to the NT group, induces cardiac hypertrophy characterised by better systolic performance accompanied by better relaxation and an improvement in the heart's mechanoadrenergic response. They also observed a slight increase in LV wall thickness and mass, and an increase in cardiomyocyte diameter. The authors emphasise that the responses depend on the intensity, duration and frequency of exercise. With regard to detraining, a sedentary period of 8 weeks after training caused the animals to return to their pre-training baseline levels.

In a study carried out in New Castle, England, Cassidy et al (2016) (Cassidy et al., 2016) used 24 male patients with type 2 diabetes mellitus and subjected them to HIIT training. They divided the patients into 2 groups: control (n=14) and HIIT (n=14). The patients in the HIIT group trained for 12 weeks, with 3 training sessions/week with exercise intensity based on the Borg scale. This training led to a slight increase in LV mass, end-diastolic volume (indicating an expansion of the LV chamber), improvement in systolic function which led to an increase in ejection volume and an increase in LV ejection fraction.

Once again, it is possible to identify the benefits of physical training on cardiac response. What's more, HIIT is being highlighted as a new and agile form of non-pharmacological intervention for treating the damage caused to the cardiovascular system and heart function.

2.4 Local Renin-Angiotensin System (RAS) and cardiovascular disease

The RAS is one of the most phylogenetically ancient hormonal systems and is therefore well conserved, especially in highly developed species (Steckelings et al., 2009). In ancient times, the RAS served to balance electrolyte homeostasis in times of salt shortage. In addition, it was and continues to be substantially involved in maintaining body circulation in situations of rapid volume loss. When overactivated, RAS contributes to many diseases, such as lesions caused by hypertension, diabetes and atherosclerosis (Hall et al., 1990).

According to the traditional view, angiotensinogen produced by the liver is converted into Angiotensin (Ang) I through the action of renin, an enzyme synthesised by cells in the juxtaglomerular apparatus of the kidneys (Hall et al., 1990). Subsequently, Ang I is cleaved by the angiotensin-converting enzyme (ACE) produced in the lungs and kidneys, thus generating Ang II, which exerts its effects by binding to two G protein-coupled receptors called angiotensin receptor type 1 (AT1) and type 2 (AT2) (Touyz and Berry, 2002).

In the last two decades, the understanding of RAAS has been amplified by the identification of new enzymes, receptors and various locally active mediators, including Ang-(1-7), Ang III (Simoes e Silva et al., 2013). These peptides are formed through the hydrolysis of Ang I or Ang II by various enzymes (Vickers et al., 2002). For example, ACE2 is a zinc metalloprotease homologous to ACE, which converts Ang II directly into Ang-(1-7) or Ang-(1-9) from the hydrolysis of Ang I (Donoghue et al., 2000; Rice et al., 2004). However, Ang-(1-7) is produced in the ovaries and kidneys, acting mainly through the action of ACE 2 on Ang II, which has 400 times more affinity for ACE2. Ang-(1-7) binds to a specific receptor called the Mas receptor, which is also a receptor coupled to transmembrane G proteins (Santos et al., 2003).

The interaction of Ang-(1-7) with the Mas receptor triggers intracellular mechanisms and functional events that, in general, oppose the actions triggered by Ang II. Ang-(1-7) produces vasodilation, inhibition of cell growth, antithrombotic, anti-inflammatory and anti-fibrotic effects (Passos-Silva et al., 2015). Therefore, it has been proposed that activation of both RAS counter-regulatory axes, ACE2/Ang-(1-7)/rMas and ACE2/Ang-(1-9)/AT2, may oppose the effects of ACE/AngII/ AT1 and prevent or reverse organ damage in experimental models of kidney and heart disease (Flores-Munoz et al., 2012).

The presence of RAS in specific organs has been demonstrated for the heart, large arteries and arterioles, kidneys and other organs, and its activation leads to structural and functional changes, independent of those caused by classical RAS. The components of this local RAS, for example, have been found in cells and tissues and some of their local functions play an important role in cellular homeostasis (De Mello, 2017). In addition, the synthesis of several components of RAS has been detected in the heart (Bader, 2002) and previous studies in pigs have indicated that up to 75% of cardiac Ang II is synthesised locally (van Kats et al., 1998). Ang II AT1 and AT2 receptors are rapidly internalised, contributing to the negative regulation of renin expression in cardiomyocytes (Hein et al., 1997).

In humans, Ang II gradients throughout the heart were increased in patients with congestive heart failure, a finding correlated with cardiac wall stress. In cases of heart failure, the local concentration of Ang II is elevated and the amount of cardiac Ang II released is related to the pathological signs of heart failure (Serneri et al., 2001). The precise role of the different components of the RAS in heart disease is not fully understood. However, transgenic mouse models have been developed to examine the role of the RAS in cardiac hypertrophy and the results have shown the presence of ventricular hypertrophy or fibrosis in some models, but not in others (Bader, 2002).

According to these studies, ventricular hypertrophy is much more dependent on haemodynamic changes than on local Ang II levels (Xu et al., 2010). Recent studies in transgenic mice have indicated that when haemodynamic conditions remain unchanged, cardiac Ang II does not alter heart size or cardiac functions (Xu et al., 2010). However, in animals with hypertension, cardiac Ang II, via the AT1 receptor, increases inflammation, oxidative stress and cell death (probably via downregulation of PI 3-kinase and Akt), thus contributing to cardiac hypertrophy and fibrosis. Studies in TG1306/1R mice have also shown that long-term overexpression of angiotensinogen in the heart leads to increased synthesis of Ang II, with consequent systolic and diastolic dysfunction and impaired excitation-contraction coupling (Domenighetti et al., 2005).

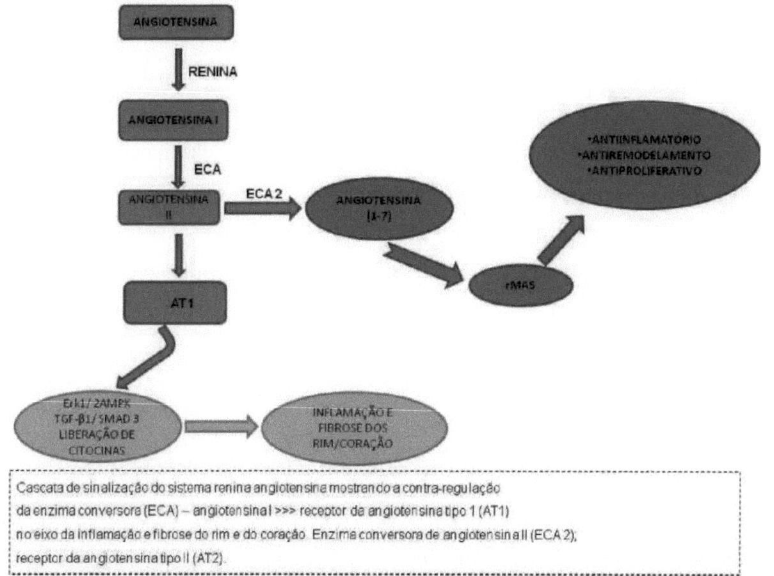

Figure 2: Renin Angiotensin System cascade. The conventional RAS, including the opposing effects of Ang II and Ang (1-7) on cardiovascular pathophysiology.

Note: adapted from Melo and collaborators Source: De Mello, 2017

2.5 Cardiovascular effects of high fructose intake

Pure, white and deadly, this is how the harmful side of sugar has been described, with its deleterious actions described many years ago, when an association between sugar consumption and coronary heart disease was established (Yudkin, 1963). Fructose, in addition to contributing half of the total content of table sugar, can also be found as a hexose in fruit and honey. More recently, sweeteners have begun to be produced from corn by isolating the starch and hydrolysing it to glucose, followed by enzymatic isomerisation of part of the glucose into fructose (Wolf et al., 2008).

The resulting mixture, known as high-fructose corn syrup, has several industrial advantages over sugar, the most important being that due to its low price, it has progressively replaced sugar consumption in North America over the last 30 years. The metabolism of fructose will be briefly outlined here. In the intestine, fructose is transported by specific transporters, GLUT5, located in the membrane of the brush border enterocyte of the small intestine (Douard and Ferraris, 2008). It reaches the hepatic portal bloodstream via another transporter, GLUT2, present in the basolateral membrane of the enterocyte; fructose then rapidly enters the hepatocyte, also via the GLUT2 transporter, without the need for ATP hydrolysis (Cheeseman, 1993; Colville et al., 1993).

Fructose is immediately phosphorylated on carbon 1, resulting in fructose-1-phosphate (P), by the enzyme fructokinase, which is highly specific for fructose. Fructose can also be phosphorylated on carbon 6, fructose-6-P, by the enzyme hexokinase, but this has a greater affinity for glucose and this reaction occurs to a lesser extent (Hallfrisch, 1990). Fructose-1-P is metabolised into two trioses-P, dihydroxyketone and glyceraldehyde-3-P, by the enzyme aldolase-B. These can follow three different paths: the glycolytic pathway, to produce energy through pyruvate; the lipogenesis pathway again, synthesising lipids and the glycogenesis pathway to produce glucose to be stored as glycogen (BARREIROS et al., 2005). At the same time, fructose inhibits the oxidation of hepatic lipids, thus favouring the re-esterification of fatty acids and the synthesis of very low density lipoprotein (*VLDL)* (Topping and Mayes, 1972).

Since fructose metabolism does not depend on insulin secretion, at least for its initial steps, and because fructose intake causes only a limited increase in glycaemia, fructose was initially proposed as a natural substitute for sucrose for diabetics. However, it quickly became apparent that a higher dietary intake of fructose has serious adverse metabolic effects in both rodents and humans (Tappy and Le, 2010). Thus, it has been recognised that high fructose intake is associated with increased plasma triglyceride concentrations, hepatic steatosis, impaired glucose tolerance and insulin resistance, and high blood pressure (Havel, 2005).

The adverse effects of fructose on glucose metabolism are closely linked to changes in lipid metabolism. In rats, intrahepatic fat content and VLDL concentrations increased after 6 weeks of eating a high-fructose diet, while intramuscular fat content increased in about 3 months. Interestingly, hepatic insulin resistance is observed shortly after switching to a high fructose diet (Bizeau and Pagliassotti, 2005). This suggests that fructose-induced insulin resistance is closely linked to ectopic lipid deposition (Unger, 2003). In addition, fructose is also known to increase plasma uric acid, which may be involved in the development of insulin resistance. In rats, fructose-induced hyperuricaemia results in the inhibition of nitric oxide synthase (NO) and it has been proposed that the inhibition of the vascular effects of NO by uric acid was involved in fructose-induced insulin resistance (Nakagawa et al., 2006).

Previous studies have shown that diets rich in simple sugars (> 20% kcal) can result in elevated triglycerides, a known risk factor for cardiovascular disease (Aeberli et al., 2011). For this reason, the American Heart Association recommends avoiding excess fructose as a nutritional mechanism for preventing hypertriglyceridaemia (Miller et al., 2011).

Variable effects related to sugar consumption and blood pressure have been reported. Johnson et al. proposed a mechanism by which fructose metabolism may result in increased levels of uric acid, which subsequently causes a reduction in endothelial nitric oxide concentrations, which is linked to hypertension (Johnson et al., 2007).

Another mechanism that explains the relationship between fructose and hypertension is that excess fructose leads to increased expression of AT1-R in rodents (Giacchetti et al., 2000); the third mechanism suggests that the presence of carbohydrates in the diet would lead to high concentrations of circulating insulin, which can act centrally to stimulate sympathetic neural activity in the heart, resulting in increased heart rate and cardiac output (Charriere et al., 2016).

2.6 Effects of a high-fat diet on cardiovascular disease

Fat, carbohydrates and proteins are the primary source of energy provided by the macronutrients consumed on a routine basis by humans. Among the macronutrients, fat contains the highest amount of energy per gram (9kcal/g). In this context, the quality, rather than the quantity, of carbohydrate and fat has become a relevant issue in nutritional origins and cardiometabolic conditions (Mozaffarian et al., 2011). Interest in the relationship between dietary fat and cardio vascular disease (CVD) arose from animal studies indicating that dietary cholesterol causes lesions, largely mediated by an increase in plasma cholesterol. Since then, the relationship between dietary fat and CVD risk has been the subject of intense research (Willett, 2012). In addition, high-fat diets, particularly those rich in saturated fatty acids, can cause low-grade systemic inflammation, insulin resistance and obesity (Erridge et al., 2007).

Consumption of a diet rich in lipids is the main risk factor for metabolic disorders linked to obesity and altered lipid metabolism and insulin sensitivity (Bessesen, 2008; Mirza, 2011), increasing the risk of developing type II diabetes (Hu et al., 2001). Adverse effects of high-fat diets on metabolic homeostasis are linked to the physiology of adipose tissue (Macotela et al., 2009). The imbalance between calorie intake and expenditure leads to adipocyte hyperplasia and hypertrophy (Funaki, 2009). In addition to the risk of developing CVD, in the last decade the focus on the composition of dietary fatty acids has been due to the induction of hepatic steatosis, which is prevalent today and correlates with an increase in free fatty acids, especially saturated ones (Leamy et al., 2013),

The "lipid hypothesis" of CVD originated with the investigations of Ancel Keys in the 1950s (Keys, 1953) and became exacerbated after his study in the 1970s. Keys claimed that there was a correlation between high dietary fat content, particularly saturated fatty acids (SFA), and both high serum total cholesterol and LDL-C. Since then, fat, and especially the consumption of SFA, has been consistently considered harmful (Keys, 1980).

It is universally accepted that trans fatty acids increase the risk of CVD, probably through their pro-inflammatory properties. Trans fats are found in natural sources, but the big villain is industrially produced trans fats, found, among other things, in fast food, bakery products, margarine and ice cream. A meta-analysis of observational studies showed that replacing 2% of energy from carbohydrates with 2% of energy from industrially produced trans fat corresponds to a 20-30% higher risk of myocardial infarction and CVD mortality (Brouwer et al., 2010).

CHAPTER 3

MATERIAL AND METHODS

3.1 Animals and experimental design

This study was approved by the local ethics committee (CEUA / 013/2016). The animals were kept in crates under suitable conditions of temperature (21 ± 2°C) and humidity (60 ± 10%), with free access to food and water. The environment was subjected to 12h light-dark cycles and air exchange (15min/h) on ventilated shelves (EcoFlo system, Allentown, USA) in accordance with the *"Guide for the use of laboratory animal care"* (NIH Publication No. 85-23, revised in 1996, USA). Sixty 12-week-old male C57BL/6 mice were randomly divided into three groups according to the diets consumed (standard diet, high-fat diet and high-fructose diet) and were monitored for 8 weeks before the HIIT training period. The standard and high-fructose diets were isocaloric, differing only in the amount of fructose, and the high-fat diet had a higher saturated fat content from lard, as seen in Table 1. The diets were prepared following the AIN-93 M guidelines (Reeves *et al.*, 1993) and were produced by PragSoluções (Jau, SP, Brazil).

After the period of induction of metabolic changes (8 weeks), the mice were randomly subdivided into six groups (n = 10 each group) for a 12-week period of high-intensity interval training (HIIT) or not, as shown below:

A) C-NT: control diet before and after HIIT and not trained for an additional twelve weeks;

B) C-T: control diet before and after HIIT and trained for an additional twelve weeks;

C) HF-NT: high-fat diet before and after HIIT and not trained for a further twelve weeks;

D) HF-T: High fat diet before and after HIIT and trained for an additional twelve weeks;

E) HFRU-NT: High fructose diet before and after HIIT and not trained for a further twelve weeks;

F) HFRU-T: High fructose diet before and after HIIT and not trained for an additional twelve weeks.

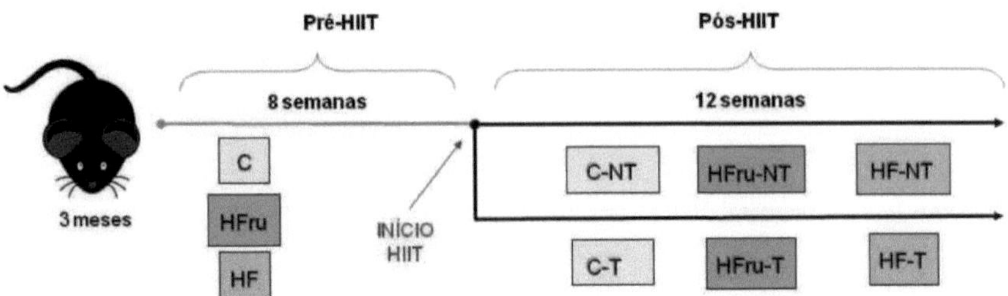

Figure 3 - Timeline illustrating the division of the groups into pre- and post-HIIT periods.

Caption: Control untrained (C-NT); Control trained (CT); High fructose diet untrained (HFRU-NT); High fructose diet trained (HFRU-T); High fat diet untrained (HF- NT); High fat diet trained (HF-T).

Source: the Author, 2017

Table 1- Diet composition

	DIETS		
Nutrients	**Control**	**HFRU**	**HF**
Casein	140,0	140,0	175,0
Maize starch	620,7	296,8	347,7
Sucrose	100,0	100,0	100,0
Lard	---	---	238,0
Fructose	---	323,85	---
Soya oil	40,0	40,0	40,0
Fibre	50,0	50,0	50,0
Vitamin mix	10,0	10,0	10,0
Mineral Mix	35,0	35,0	35,0
Cysteine	1,8	1,8	1,8
Hill	2,5	2,5	2,5
Antioxidant	0,008	0,008	0,008
Total (g)	**1000**	**1000**	**1000**
Energy	**3802,8**	**3802,8**	**5000**
Carbohydrate (%, energy)	**76**	**76**	**36**
Fructose (%, energy)	---	**34**	---
Protein (%, energy)	**14**	**14**	**14**
Lipids (%, energy)	**10**	**10**	**50**

Legend: HFRU - high in fructose; HF - high in fat Source: Pimenta et al (2015); Schultz et al (2015)

3.1.1 Exercise Protocol

All mice assigned to the HIIT exercise training group were familiarised with a treadmill (Treadmill Control LE 8710 - Panlab Havard Apparatus) on 3 occasions (10m/min, 0% grade, 10-15 min). The exercise protocols were developed and adapted based on previous rodent training experience at our institution (Pimenta et al., 2015). To assess improvements in exercise performance with training, an exercise capacity test was performed at the beginning, middle and end, before each stage of the training protocol, to observe and adjust the intensity according to the animals' evolution. The exercise test was based on speed increments every 2 minutes from a minimum speed of 10m/min (Borges et al., 2014). The HIIT groups also trained for 3 days/week for 12 weeks with variations in intensity during training in 3-minute cycles, 1 minute of low intensity (30%) and 2 minutes of high intensity (80%) until exhaustion, total daily duration in minutes according to the stress test. Exhaustion was defined as the point at which, instead of running on the treadmill, the mice remained on the shock devices, which are designed to encourage them to perform an additional 10s of training. In the first two weeks, 80 per cent of the maximum exercise intensity was used, in the third week the value was increased to 85 per cent and in the fourth week to 90 per cent, when the last stress test was carried out. During the HIIT period, the untrained groups remained in their cages with water and diets ad libitum.

3.1.2 Oral glucose tolerance test (OGTT) and quantitative insulin sensitivity index (QUICKI)

After 6 hours of fasting, the mice were given 1.0 g/kg of glucose via orogastric gavage for the TOTG. Serial blood samples were obtained via a tail notch only once and at times 0, 15, 30, 60, 90, 120 minutes, and the glucose level was assessed using Accu-Chek (Accu-Chek, Roche Diagnostic, Manheim, Germany). The glucose reading at each time point was used to calculate the area under the curve and was compared between the experimental groups. The Quantitative Insulin Sensitivity Check Index (QUICKI) was used to measure insulin sensitivity (Katz et al., 2000), determined by the following mathematical equation:

QUICKI= 1 /[log (fasting insulin) + log (fasting glycaemia)])

3.1.3 Indirect Calorimetry

Two gas exchange determinations (oxygen [O2] consumption and carbon dioxide [CO2] production) were carried out using open-circuit respirometry equipment (Metabolism Oxylet System, Panlab / Harvard, Barcelona, Spain). Measurements began 24 h after the last exercise session to avoid any post-exercise effects. The animals were placed individually for 72 h (with a 48 h period of acclimatisation to the cages, the last 24 h being considered for analysis) in an insulated polycarbonate cage, with free access to food and water. The system makes it possible to monitor the amount of oxygen volume (V:O2) and the volume of carbon dioxide (V:CO2) produced sequentially for each cage (every 15 min, for 3 min in each cage), where the respiratory quotient (RQ) is given by the well-known equation: RQ = V:CO2 / V:O2 and energy expenditure (EE) by the Weir equation: EE =

[3,815 + (1,232 x RQ)] x VO2 x 1.44; With the units in kcal/day kg^ 0.75. Calibration of the O2 and CO2 analysers was carried out before and after each experiment, and was determined as O2 = 20%, CO2 = 0% and O2 = 50%, CO2 = 2%.

3.1.4 Systolic blood pressure (SBP)

SBP measurements were always taken in the morning at the same time, using a calibrated tail gauging system (LE 5002 storage pressure meter - Panlab Havard Apparatus). The mice were kept warm using a heating pad and were acclimatised and trained at least 3 weeks before the actual measurements were taken to be used for data analysis. The average of 3 measurements per animal was used for the final blood pressure measurement, which was used to analyse the data.

3.1.5 Euthanasia

The animals were fasted for 6 hours on the day of euthanasia. They were then deeply anaesthetised (intraperitoneal sodium pentobarbital, 100 mg / kg-1) and blood samples were obtained quickly by cardiac puncture.

3.1.6 Heart analysis

The heart was removed, the left ventricle (LV) was dissected and its mass was measured using the Scherle method. The length of the left tibia was also measured to normalise the LV mass (Yin et al., 1982). The LV was quickly fixed by immersion in fixative (1.27 mol/L formaldehyde in 0.1 M phosphate buffer, pH 7.2) for 48 h at room temperature. The LV was then embedded in Paraplast plus (Sigma-Aldrich Chemical Co), sectioned at 5 pm and stained with haematoxylin and eosin. The compact portion of the LV wall thickness was measured in the valve plane. The analysis used a Leica DMRBE microscope (Wetzlar, Germany) and Lumenera Infinity 1-5c (Ottawa, Canada) with Image Pro Plus v. 7.01 software for Windows (Media Cybernetics, Silver Spring, USA).

3.1.7 Biochemical analysis

Plasma was separated by centrifugation (120g for 15 min) at room temperature and used to measure total cholesterol, triglycerides using a semi-automatic spectrophotometer and appropriate commercial kits (Bioclin, Quibasa, Belo Horizonte, MG). Plasma insulin concentration was analysed in duplicate using the enzyme-linked immunosorbent assay kit (Rat/Mouse Insulin ELISA Cat. # EZRMI-13K, Millipore, Missouri, USA), using the TPREADER Thermoplate equipment (Bio Tek Instruments, Inc Highland Park, USA).

3.1.8 Uric acid

Urine was collected over 48 hours using metabolic cages for individualised animals; it was stored at -20°C until the time of analysis. A 4 ml aliquot of urine was used for the enzymatic determination of uric acid by enzymatic colorimetric method using an automated spectrophotometer and commercial kits. The intensity of the cherry colour formed is directly proportional to the concentration of uric acid in the sample (Bioclin II System, Quibasa Ltda, Belo Horizonte, MG, Brazil).

3.1.9 Western blot

The LV myocardial tissue samples were frozen in liquid nitrogen and stored at -80°C. Total proteins were extracted in a homogenising buffer with protease and phosphatase inhibitors. The proteins were separated by polyacrylamide gel electrophoresis, transferred to a nitrocellulose membrane, blocked at room temperature for 2h, and incubated overnight at 4°C with the primary antibodies,Renin (anti-mouse, SC137252; Santa Cruz Biotechnology; 1:1000), ACE (anti-mouse, ab11734; Abcam; 1:500), AT1R (anti-rabbit, SC579; Santa Cruz Biotechnology; 1:1000) and AT2R (anti-goat, SC48452; Santa Cruz Biotechnology; 1:1000) . After incubation with the primary antibody, the transfers were incubated with appropriate secondary antibodies for 1 hour and then incubated with Clarity Western ECL Substate detection reagents. Images of the transfers were obtained using the ChemiDoc Molecular Imaging System (Bio-Rad, CA). Beta-actin (anti-mouse, Sigma Aldrich; 1:5000) was used as a loading control for proteins.

3.2 RT-qPCR

RNA was isolated from ventricular samples using Trizol reagent (Invitrogen, CA). Total RNA was treated with Dnase (Invitrogen, CA, USA) and complementary first-strand DNA was synthesised using oligonucleotides and RT Superscript III (Invitrogen, CA). Real-time PCR was carried out using the Bio-Rad CFX96 thermal cycler and the SYBRGreen mix (Invitrogen, CA). The endogenous control beta-actin was used to normalise the expression of the selected genes. RT-qPCR efficiencies for the target gene and the control were approximately equal and were calculated using cDNA dilution sequences. The RT-qPCR reactions were conducted as follows: after a pre-denaturation and polymerase activation programme (4 min at 95°C), 44 cycles, each consisting of 95°C for 10 s and 60°C for 15 s were followed by heating (60 to 95°C with a heating rate of 0.1°C/s). Negative controls consisted of wells in which the cDNA was replaced with deionised water. The relative expression ratio (RQ) of the mRNA was calculated using the 2-AACt equation, in which -ACT expresses the difference between the number of cycles (CT) of the target genes and the endogenous control. The sequences of the direct and reverse primers used for amplification are described below:

a) Renina,(5'-3')ACCTTGCTTGTGGGATTCAC,
(3'-5') CTGATCCGTAGTGGATGGT;
b) ACE,(5'-3') GTGGCTGGAAGAGCAGAATC,
(3'-5') GCCTTGGCTTCATCAGTCTC;
c) ACE2,(5'- 3') CAACAGAAGCCAGACAACA ,
(3'-5') GCCTTGGCTTCATCAGTCTC;
d) AT1R,(5'-3')CCCTGGCTGACTTATGCTTT,
(3'-5') ACATAGGTGATTGCCGAAGG;
e) AT2R,(5'-3')GAAGCTCCGCAGTGTGTTTA,
(3'-5') TGGCTAGGCTGATTACATGC;
f) MASr,(5'-3') TTCTCCACCATCAACAGCAG ,
(3'-5') CCTGGGTTGCATTTCATCTT;
g) B-ACTIN,(5'-3')TGTTACCAACTGGGACGACA,
(3'-5') GGGGTGTTGAAGGTCTCAAA

3.3 Statistical Analysis

The data was tested for normality and homoscedasticity of variance and expressed as mean and standard deviation (SD). The differences between the groups in the pre-HIIT period (groups C, HF and HFRU) were tested using one-way ANOVA followed by the Holm-Sidak post-test. In the post-HIIT period, two-way ANOVA followed by Holm-Sidak post-test was used to analyse the contribution of each factor individually, diet, HIIT and the interaction between these factors (GraphPad Prism version 7.0 for Windows, GraphPad Software, La Jolla, CA, USA). P-values <0.05 were accepted as statistically significant.

CHAPTER 4

RESULTS

4.1 Pre-HIIT

4.1.1 Body Mass (BM)

The animals started the experiment with no significant difference in body mass. Diets rich in fat or high fructose were administered for 8 weeks before starting the HIIT protocol, in order to cause changes in body mass and consequent metabolic changes. Thus, at the end of 8 weeks of administration of the respective diets, the BM of the HF-NT group was 15% higher than the C-NT group and 22% higher compared to the HFRU group (P <0.0001). Results in table 2.

4.1.2 Oral Glucose Tolerance Test - TOTG

TOTG was performed to assess glucose homeostasis and the corresponding area under the curve (AUC) was analysed. Significant glucose intolerance was observed in the HF-NT and HFRU-NT groups. The HF-NT group showed an AUC approximately 21% higher than the C-NT group (P<0.0010), similarly the HFRU-NT group also showed an increase in AUC approximately 9% higher compared to the C-NT group (P<0.0375); thus, the HF-NT and HFRU-NT groups showed glucose intolerance before starting HIIT (Figure 4).

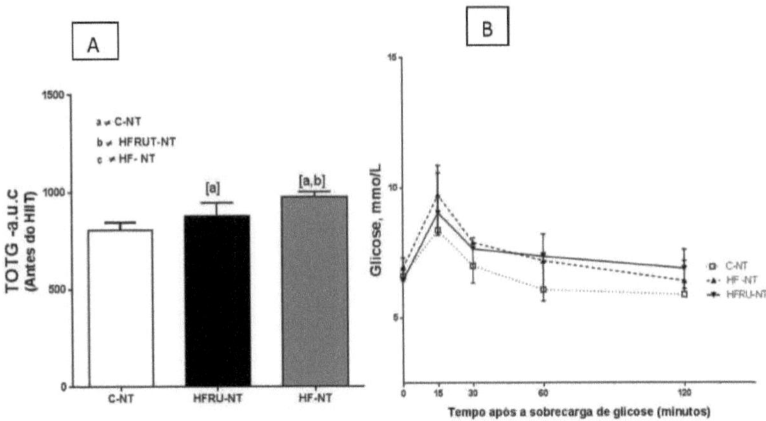

Figure 4- TOTG 8 weeks before starting HIIT.

Legend: untrained control (C-NT); untrained high fructose diet (HFRU-NT); untrained high fat diet (HF-NRT)
Note: in A, graph of the area under the curve. In B, evolution of glycaemia during the 120 minutes after glucose overload. Differences are shown according to the internal legend (P <0.05, one-way ANOVA and Holm-Sidak post-test), when: [a] different from C-NT; [b] different from HFRU-NT, [c] different from HF-NT.
Source: the Author, 2017

4.2 Post-HIIT

HIIT reduces body mass, increases energy expenditure and improves glucose tolerance

4.2.1 Body mass (BM)

At the end of the 12-week training session, the BM of the HF-T group showed a significant reduction of - 6.5% when compared to the HF-NT group (P = 0.03), similar differences were not observed in the C-NT and HFRU-NT groups compared to their respective counterparts, since these groups showed no significant increase in body mass (Table 3). Only diet independently influenced final body mass, accounting for 79.28 per cent of its total variance (two-way ANOVA, P<0.0001). In addition, there was a significant interaction between diet and HIIT in determining this parameter (two-way ANOVA, P = 0.0380).

4.2.2 Indirect calorimetry

As for the respiratory quotient (RQ), the HFRU-NT group had a higher RQ than the other untrained groups, C-NT and HF-NT (P <0.0001); and after HIIT, the trained groups showed no significant changes. With regard to energy expenditure (EE), the HF-T group had a significantly higher EE compared to the HF-NT group (P = 0.03), which helps to justify the reduction in body mass observed in this group. Table 4.

Table 2 - indirect calorimetry.

Date	C-NT	C-T	HFRU-NT	HFRU-T	HF-NT	HF-T
Quotient respiratory	0,92±0,09	0,95±0,09[a]	0,98±0,08[a]	1_{1} 09±0_{1} 09	0.83±0.02[a,b]	0,89±0,08[c]
Spending energetic (kcal/ day/kg/^A 0.75)	147±17,9	155±18	158±14.4[a]	164,9±18	184,9±20[a]	195,1±11[c]

Caption: Caption: untrained control (C-NT); trained control (C-T); untrained high-fructose diet (HFRU-NT); trained high-fructose diet (HFRU-T); untrained high-fat diet (HF-NRT); trained high-fat diet.
Note: Data are presented as mean and standard deviation of the mean. In cases marked P <0.05, (one-way ANOVA with Holm-Sidak post-test), then: [a] different from C-NT; [b] HFRU-NT, [c] HF-NT.
Source: the Author, 2017

4.2.3 Oral glucose tolerance test - area under the curve (AUC)

In general, the glucose values and AUC for TOTG were significantly higher in the HF-NT and HFRU-NT groups than in their respective counterparts. The AUC of the HF-T group decreased significantly compared to the HF-NT group (P = 0.04), similarly, the HFRU-T group showed a significant reduction in AUC when compared to the HFRU-NT (P = 0.005). The two-way ANOVA analysis revealed that both diet and HIIT exerted isolated influences on the AUC for TOTG, without interacting significantly. HIIT accounted for 39.13 per cent of the total variance with regard to this parameter (P < 0.0001), while diet accounted for 27.38 per cent of its total variance (P = 0.0003), figure 5.

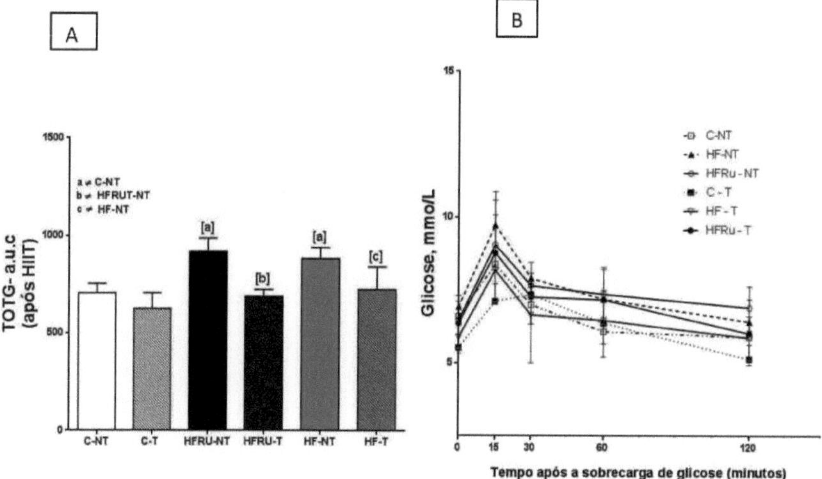

Figure 5- TOTG 12 weeks after HIIT.

Legend: untrained control (C-NT); trained control (C-T); untrained high fructose diet (HFRU-NT); trained high fructose diet (HFRU-T); untrained high fat diet (HF- NRT); trained high fat diet.
Note: in A, graph of the area under the curve. In B, evolution of glycaemia during the 120 minutes after glucose overload.Differences are shown according to the internal legend (P <0.05, one-way ANOVA and Holm-Sidak post-test), when: [a] different from C-NT; [b] different from HFRU-NT, [c] different from HF-NT.
Source: the Author, 2017

HIIT improves insulin sensitivity in groups fed a high-fat or high-fructose diet

4.2.4 Quantitative insulin sensitivity test index (QUICKI)

Insulin resistance has been described as one of the factors in cardiovascular complications. As expected, animals fed a high-fat or high-fructose diet developed reduced insulin sensitivity, which is characterised by hyperinsulinemia and impaired fasting glucose. Notably, the HF-NT group showed reduced insulin sensitivity associated with obesity, while the HFRU-NT group developed a similar condition, but with no association with obesity. However, after HIIT, insulin sensitivity was better in the HF-T,+ 14% group when compared to the HF-NT group (P <0.0001). Interestingly, HIIT was able to improve insulin sensitivity in the HFRU-T group, + 7%, compared to the HFRU-NT group (P <0.0001), table 3.

HIIT effectiveness and biochemical changes

Data shown in table 3

4.2.5 Total cholesterol (TC)

Total cholesterol was significantly higher in the HFRU-NT and HF-NT groups compared to the C-NT group (P = 0.0003 and 0.0005, respectively). Regarding the effect of HIIT on this parameter, a significant reduction in TC of around 18.5 per cent was observed in the HF-T group compared to the HF-NT group (P = 0.02). However, no similar reduction was observed in the HFRU-T group when compared to the HFRU-NT group (P = 0.74).

4.2.6 Triglycerides

As for triglycerides, both the high-fat diet and the high-fructose diet were able to alter this parameter in the HF-NT and HFRU- NT groups compared to the C-NT group (P <0.0001; P = 0.007, respectively). On the other hand, the trained HF-T and HFRU-T groups showed no significant reduction compared to their counterparts.

4.2.7 Uric acid

High fructose intake is known to increase the risk of hyperuricaemia, so the HFRU-NT group showed a significant increase in uric acid compared to the C-NT group (P=0.004); however, when compared to its HFRU-T counterpart, no significant reduction was observed (P=0.83). With regard to high fat consumption, the HF-T group showed a -11 per cent reduction in uric acid when compared to the HF-NT group (P = 0.02). Only diet exerted a single significant influence on total cholesterol and uric acid levels (two-factor ANOVA, P = 0.0008 and P <0.0001).

HIIT is unable to reduce SBP after consuming a high-fructose diet

4.2.8 Systolic blood pressure - starts

The groups started the experiment with no difference in systolic blood pressure. After 8 weeks of eating the respective experimental diets, the HF-NT group showed a significant increase in SBP of around +16.8% (P = 0.0002) and HFRU- NT of +21% compared to the C-NT group (P = 0.0001), table 3.

4.2.9 Systolic blood pressure - final

SBP at the end of the training period and the concomitant intake of HFRU and HF diets behaved differently at the end of the experiment. Firstly, HFRU-NT and HF-NT had an SBP about + 22% higher compared to the C-NT group (P <0.0001). However, after training, the HF-T group showed a significant reduction of - 6.5% (P = 0.04) compared to its HF-NT counterpart. On the other hand, the HFRU-T group showed a reduction of around -4 per cent in SBP, although this was not statistically significant (P=0.39). Both diet and HIIT significantly influenced systolic blood pressure values (two-way ANOVA, P < 0.0001 for diet and P = 0.0137 for HIIT). However, the HFRU diet exerted a more significant influence on this parameter, since it accounted for 77.21 per cent of the total variance, while only 4.5 per cent of the total variance was associated with HIIT, table 3.

HIIT improves left ventricular hypertrophy

4.3 Left ventricular (LV) mass

The morphological changes were also microscopic. Consumption of the HF and HFRU diets induced cardiac hypertrophy after 20 weeks of dietary intake, demonstrating the ability of the diets to initiate the development of cardiac hypertrophy. The LV masses of the HF-NT and HFRU-NT groups were significantly higher when compared to the C-NT group (+ 27% and + 13%, P = 0.0001 and 0.016, respectively); the HF-NT group had a higher LV mass than the other untrained groups. However, after HIIT, there was a significant reduction in LV mass in the HF-T group compared to the HF-NT group (P = 0.016) of around -8.5%. The HFRU-T group had a significant reduction of around -13% in LV mass when compared to its untrained counterpart (P = 0.01), table 3.

4.3.1 Left ventricular wall thickness

Corroborating with LV mass, the HF-NT and HFRU-NT groups had greater left ventricular wall thickness than the C-NT group (+ 10% and 11%, respectively; P<0.0001). The cardiac hypertrophy observed in the experimental model of this study may be due to the deposition of fat in the myocardium. Comparing the HF-T group with the HF-NT group, LV wall thickness after HIIT was -6.5% lower (P = 0.02). The HFRU-T group also showed a reduction in LV wall thickness, around -6% compared to the HFRU-NT group (P = 0.030). According to the two-factor ANOVA, diet had a significant influence on the LV/tibia ratio (P<0.0001), and on LV thickness (P=0.0043). There was a significant interaction between diet and HIIT only in relation to LV/tibia (P = 0.002), figure 6.

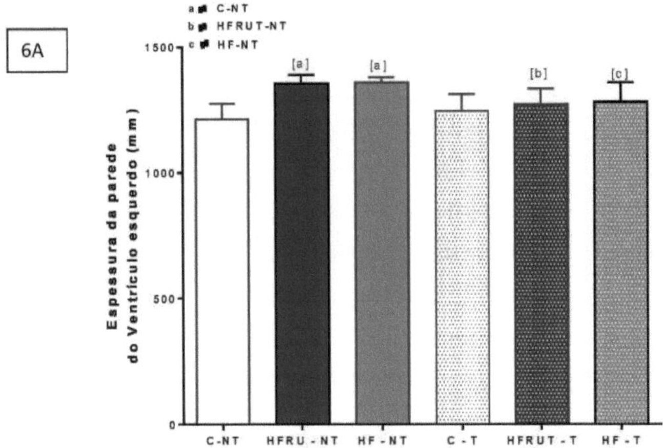

Figure 6- Left ventricular wall thickness (mm)

Legend: untrained control (C-NT); trained control (C-T); untrained high fructose diet (HFRU-NT); trained high fructose diet (HFRU-T); untrained high fat diet (HF- NRT); trained high fat diet.

Note: differences are shown in the bars according to the internal legend (P <0.05, one-way ANOVA and Holm-Sidak post-test), when: [a] different from C-NT; [b] different from HFRU-NT, [c] different from HF-NT.

Source: the Author, 2017

6B

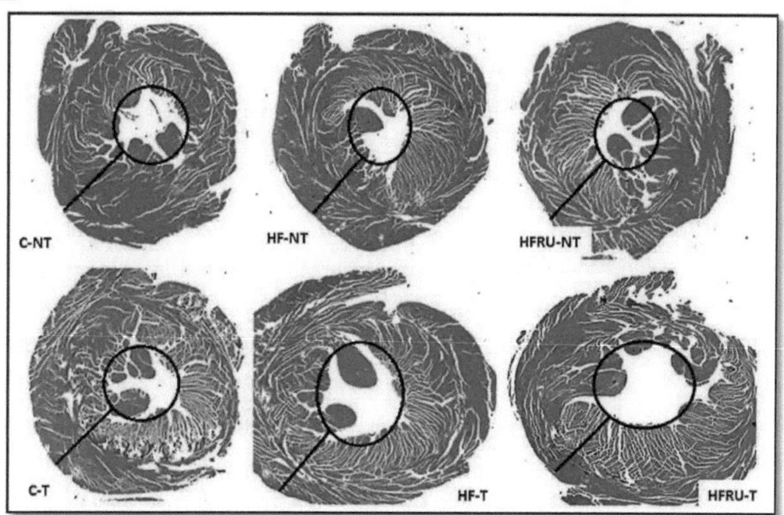

Legend: untrained control (C-NT); trained control (C-T); untrained high fructose diet (HFRU-NT); trained high fructose diet (HFRU-T); untrained high fat diet (HF- NRT); trained high fat diet.

Note: photomicrograph of the left ventricle: caudal cross-sectional view in the valve plane showing the differences in wall thickness, all at the same magnification (x4).

Source: the Author, 2017

Table 3 - Parameters: Body mass, left ventricular mass, blood pressure, triglycerides, total cholesterol, uric acid, QUICKI.

Data	C-NT	C-T	HFRU-NT	HFRU-T	HF-NT	HF-T
Initial body mass (before HIIT) (g)	27,2±1,8	-	26,3±1,2	-	31.3±1.4[a,b]	-
Final body mass (after HIIT) (g)	28,6±1,7	29,6±1,4	26,5 ±0,5	26,8±0,8	31.3±1.4[a,b]	32,1±1,0[c]
Left ventricle mass/tibia length	0,037±0,002	0,041±0,002	0,043±0,003[a]	0,037±0,002[b]	0.047±0.001[a,b]	0,043±0,002[c]
Systolic blood pressure before HIIT (mm/Hg)	124 ±4,5	-	151±5,3	-	145±6,5	-
Systolic blood pressure after HIIT (mm/Hg)	126±3,1	123±4,7	155±1,2[a]	149±4,7	152±3,8[a]	144±3,01[c]
QUICKI	0,30±0,003	0,30±0,004	0,28±0,001[a]	0,30±0,006[b]	0,27±0,005[a]	0,31±0,003[c]
Total cholesterol (mg/dl)	59,8±7,1	65,8±7,6	81,6±9,1[a]	76,3±4,4	80,5±6,04[a]	65,8±6,6[c]
Triglycerides (mg/dl)	41,2±4,8	46,8±1,4	61,8±8,7[a]	59,8±6,3	55,2±2,5[a]	55,6±5,8
Uric acid (mg/dl)	22,6±,6	22,9±0,6	25,4±0,4[a]	26,3±0,9	23,5±0,5[b]	22,4±0,5

HIIT induces changes in the protein and gene expression of the renin-angiotensin system in the left ventricle

4.4 Western blot

All Western blot data was normalised using beta-actin.

The renin-angiotensin system (RAS) is related to the physiological adaptations induced by exercise, and the cardiovascular RAS can be affected by diet. Regarding renin expression (figure 7), it was significantly increased in the HF-NT and HFRU-NT groups compared to the C-NT group (+46%, + 54%, P=0.001, respectively). After HIIT, the HF-T group showed a significant reduction in renin expression when compared to the HF-NT (-21%, P=0.009), in contrast, the HFRU-T group showed a reduction but no significant difference when compared to the HFRU-NT (P=0.18). Renin protein expression was predominantly influenced by diet (40.25% of the total variance, P <0.0001), but HIIT and the interaction between both factors (diet and exercise) also significantly influenced the results (two-factor ANOVA, P = 0.0288). As for angiotensin-converting enzyme (ACE), the HF-NT group showed higher ACE protein expression (figure 8) than the HF-T group (+ 23%, P = 0.009). Similarly, ACE protein expression was higher in the HFRU-NT group than in the HFRU-T group (+25%, P=0.001). In addition, both HF-NT and HFRU-NT showed higher ACE expression than the C-NT group (P = 0.02 and 0.002, respectively). ACE protein expression was only influenced by diet and HIIT independently (two-way ANOVA, P <0.0001). Angiotensin receptors play an important role in the response to metabolic changes. Thus, an increased expression of AT1R (figure 9) was observed in the HF-NT and HFRU-NT groups when compared to the C-NT group (P <0.0001). AT1R expression was reduced in the HF-T group compared to the HF-NT group (-60 per cent, P=0.004). Conversely, AT1R expression in the HFRU-T group was not significantly reduced compared to HFRU-NT (P= 0.07).

AT2R expression (figure 10) was reduced in all HIIT groups when compared to their counterparts. HF-T showed reduced AT2R expression when compared to HF-NT (P = 0.004) as did the HFRU-T group compared to HFRU-NT (P = 0.003). AT1R and AT2R protein expression was also influenced by diet and HIIT (two-way ANOVA, P < 0.0001), as well as being influenced by a significant interaction between both factors (P = 0.0173). It is worth noting that AT1R was predominantly influenced by an interaction between diet and HIIT (46.38 per cent of the total variance), while AT2R was predominantly influenced by HIIT (83.46 per cent of the total variance).

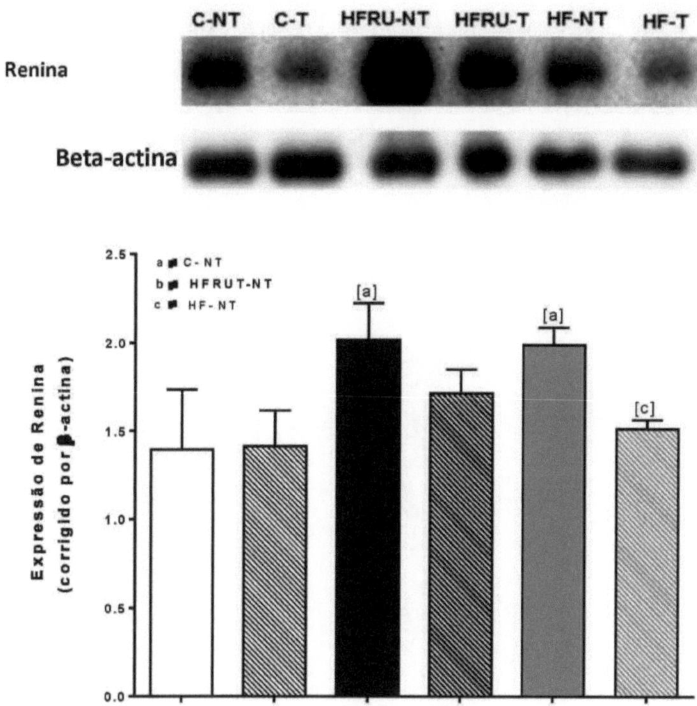

Figure 7-Renin protein expression, corrected for beta-actin

Legend: untrained control (C-NT); trained control (C-T); untrained high fructose diet (HFRU-NT); trained high fructose diet (HFRU-T); untrained high fat diet (HF- NRT); trained high fat diet.

Note: data are presented as mean ± SD, significant differences between groups are indicated by symbols P <0.05, one-way ANOVA and Holm-Sidak post-test: [a] compared to C-NT; [b] different from HFRU-NT, [c] different from HF-NT.

Source: the Author, 2017

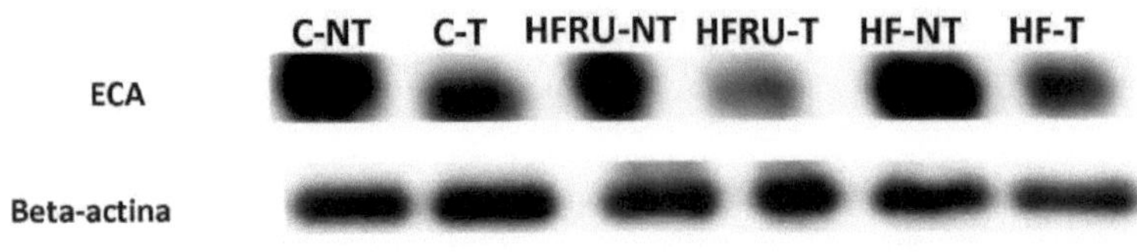

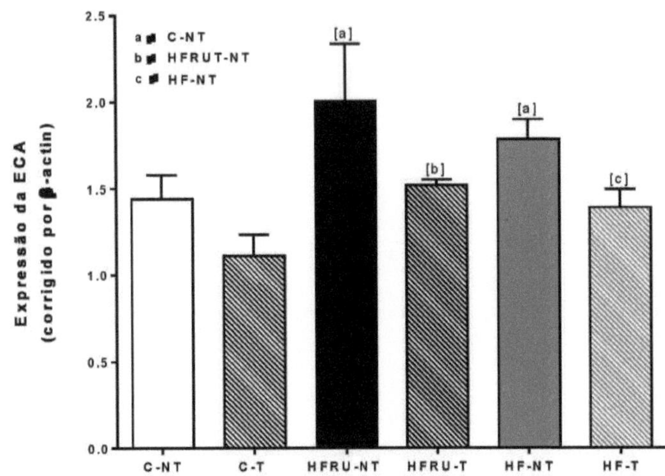

Figura 8 - ACE protein expression, corrected by beta-actin.

Legend: untrained control (C-NT); trained control (C-T); untrained high fructose diet (HFRU-NT); trained high fructose diet (HFRU-T); untrained high fat diet (HF- NRT); trained high fat diet.

Note: data are presented as mean ± SD, significant differences between groups are indicated by symbols P <0.05, one-way ANOVA and Holm-Sidak post-test: [a] compared to C-NT; [b] different from HFRU-NT, [c] different from HF-NT.

Source: the Author, 2017

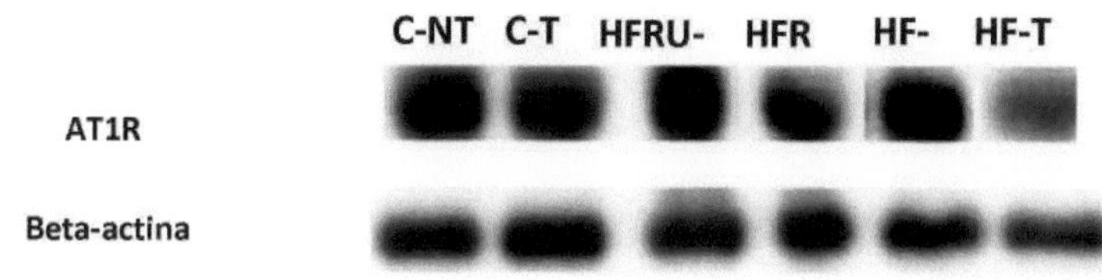

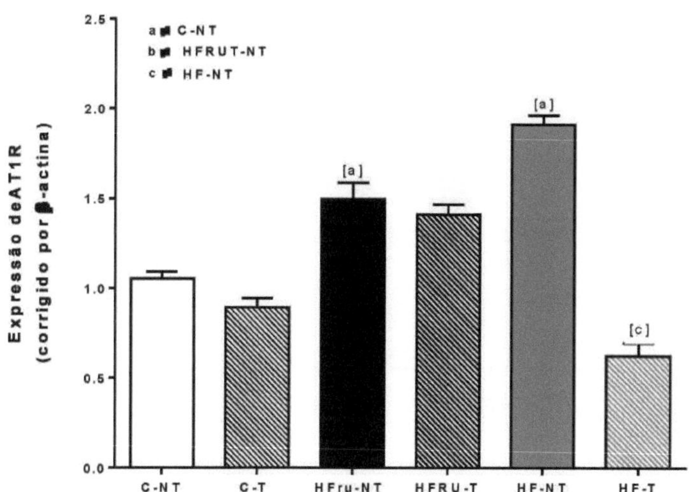

Figura 9 - AT1R protein expression, corrected by beta-actin.

Caption: untrained control (C-NT); trained control (C-T); untrained high fructose diet (HFRU-NT); trained high fructose diet (HFRU-T); untrained high fat diet (HF- NRT); trained high fat diet.

Note: Data are presented as mean ± SD, significant differences between groups are indicated by symbols $P < 0.05$, one-way ANOVA and Holm-Sidak post-test: [a] compared to C-NT; [b] different from HFRU-NT, [c] different from HF-NT.

Author: the Author, 2017

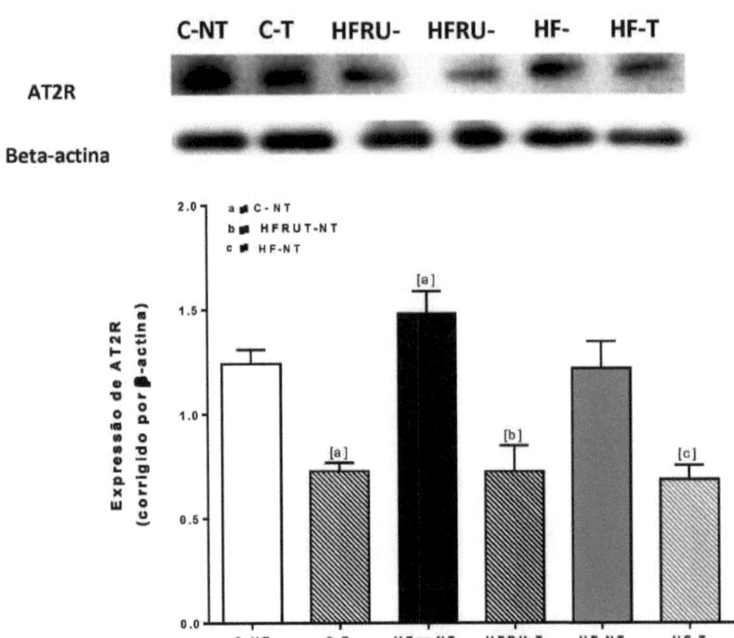

Figura 10 - AT2R protein expression, corrected by beta-actin.

Caption: untrained control (C-NT); trained control (C-T); untrained high fructose diet (HFRU-NT); trained high fructose diet (HFRU-T); untrained high fat diet (HF- NRT); trained high fat diet.

Note: Data are presented as mean ± SD, significant differences between groups are indicated by symbols. P <0.05, one-way ANOVA and Holm-Sidak post-test: [a] compared to C-NT; [b] different from HFRU-NT, [c] different from HF-NT.

Source: the Author, 2017

4.5 mRNA analysis - RT-qPCR

Changes in the mRNA of the angiotensin-renin system in the left ventricle were measured by RT-qPCR. The results of SRA axis gene expression were similar to those described above in WB. Renin gene expression (figure 11) was reduced in the HF-T group compared to the HF-NT group (P = 0.003), a similar reduction was observed in HFRU-T compared to its counterpart (P = 0.02). Renin was significantly influenced by diet (P= 0.0186), and also by HIIT (P = 0.0002) and the significant interaction between the two factors (P= 0.0005 and P <0.0001).

High-fat and high-fructose diets significantly up-regulated AT1R gene expression (figure 12) in the HF-NT and HFRU-NT groups compared to their counterparts. On the other hand, AT1R gene expression after HIIT was reduced in the HF-T group when compared to the HF-NT (P = 0.004). Similar to WB, HFRU-T also showed no significant reduction in AT1R gene expression compared to HFRU-NT (P = 0.86).

In addition, AT2R gene expression (figure 13) showed reduced expression after HIIT in the HF-T group compared to the HF-NT group (P=0.004). The HFRU-T group also showed lower AT2R gene expression than the HFRu-T group (P=0.003). AT1R gene expression was influenced by diet, HIIT and their interaction (two-way ANOVA, P <0.0001). It is worth noting that HIIT accounted for 54.45% of the total variance in AT1R gene expression. Similarly, AT2R gene expression was influenced by diet and HIIT (two-factor ANOVA, P <0.0001) and by an interaction between both factors (P = 0.0033).

The HIIT groups exhibited reduced ACE gene expression (figure14), as observed in the HF-T group compared to the HF-NT group (P = 0.0005). Similarly, ACE gene expression in the HFRU-T group was lower when compared to the HFRU-NT group (P = 0.02). On the other hand, with regard to ACE2 gene expression (figure 15), an important enzyme in the ACE2/ANG1-7/rMAS axis, this enzyme was markedly increased in the HF-T group compared to the HF-NT group (P = 0.007), as well as in the HFRU-T group compared to the HFRU-NT group (P = 0.04), demonstrating the beneficial role of HIIT on this enzyme.

important enzyme. ACE and ACE2 gene expression was significantly influenced by diet (two-factor ANOVA, P<0.0001 for both), HIIT (P <0.0001 and P = 0.0008) and the significant interaction between both factors (P <0.0001 and P <0.0125). Of note, ACE2 gene expression was predominantly influenced by diet (46.54% of the total variance) and HIIT (16.04% of the total variance).

An important part of the RAS axis is the Mas receptor (rMAS), which has been described as a functional receptor for the cardioprotective fragment of RAS. Our results showed that after HIIT, there was a marked increase in rMAS gene expression in the HF-T group compared to the HF-NT group (P = 0.0004) and a milder but significant increase in rMAS gene expression in the HFRU-T group compared to the HFRU-NT group (P = 0.02). Only diet and HIIT, as isolated factors, significantly influenced rMAS gene expression, with no significant interaction between them (two-factor ANOVA, P <0.0001, figure 16).

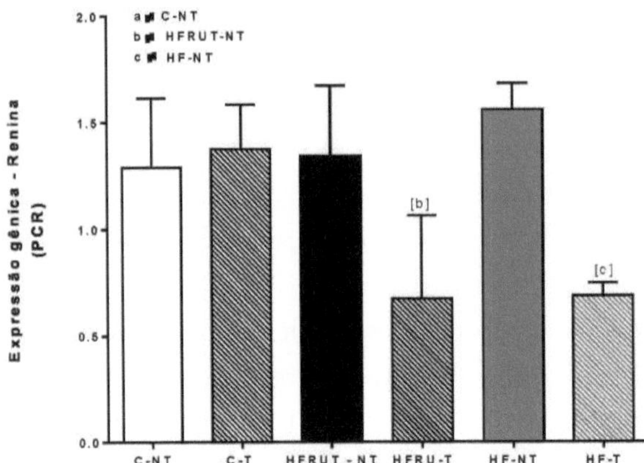

Figura 11- Renin gene expression.

Caption: untrained control (C-NT); trained control (C-T); untrained high fructose diet (HFRU-NT); trained high fructose diet (HFRU-T); untrained high fat diet (HF- NRT); trained high fat diet.
Note: data are presented as means ± SD, significant differences between groups are indicated by symbols above the bars; P <0.05, one-way ANOVA and Holm-Sidak post-test: [a] compared with C-NT; [b] different from HFRU-NT, [c] different from HF-NT.
Source: the author, 2017.

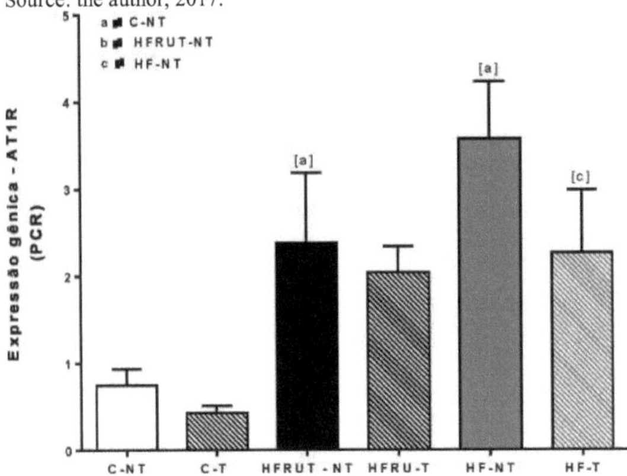

Figura 12- AT1R gene expression.

Caption: untrained control (C-NT); trained control (C-T); untrained high fructose diet (HFRU-NT); trained high fructose diet (HFRU-T); untrained high fat diet (HF- NRT); trained high fat diet.
Note: Data are presented as means ± SD, significant differences between groups are indicated by symbols above the bars; P <0.05, one-way ANOVA and Holm-Sidak post-test: [a] compared to C-NT; [b] different from HFRU-NT, [c] different from HF-NT.
Source: the Author, 2017

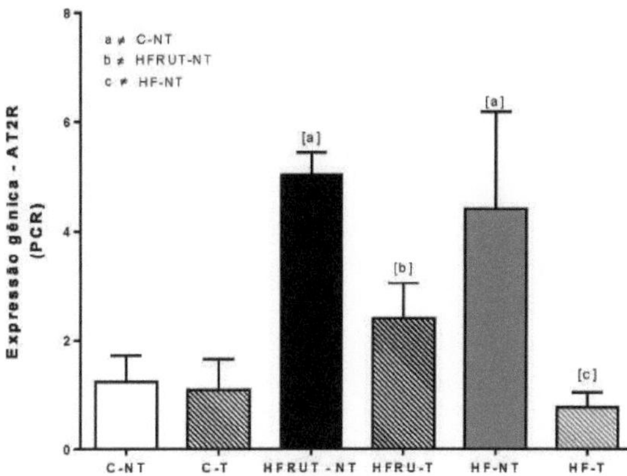

Figura 13- AT2R gene expression.

Legend: untrained control (C-NT); trained control (C-T); untrained high fructose diet (HFRU-NT); trained high fructose diet (HFRU-T); untrained high fat diet (HF- NRT); trained high fat diet.

Note: data are presented as means ± SD, significant differences between groups are indicated by symbols above the bars; P <0.05, one-way ANOVA and Holm-Sidak post-test: [a] compared with C-NT; [b] different from HFRU-NT, [c] different from HF-NT.

Source: the author, 2017

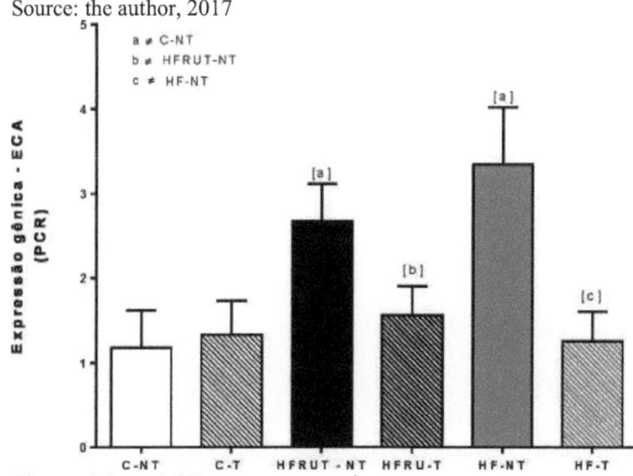

Figura 14- ACE gene expression.

Legend: untrained control (C-NT); trained control (C-T); untrained high fructose diet (HFRU-NT); trained high fructose diet (HFRU-T); untrained high fat diet (HF- NRT); trained high fat diet.

Note: data are presented as means ± SD, significant differences between groups are indicated by symbols above the bars; P <0.05, one-way ANOVA and Holm-Sidak post-test: [a] compared with C-NT; [b] different from HFRU-NT, [c] different from HF-NT.

Source: the author, 2017.

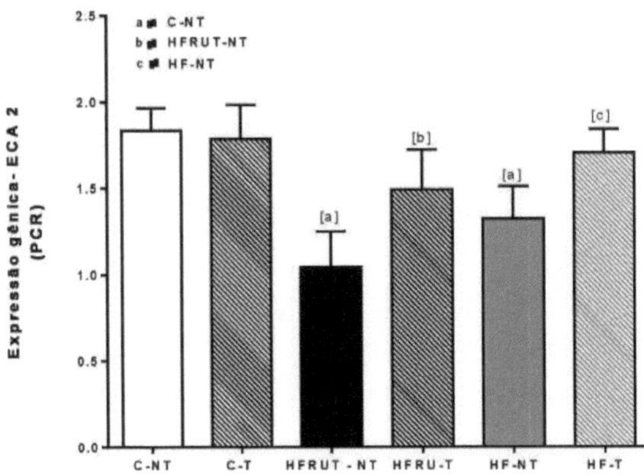

Figura 15- ACE2 gene expression.

Legend: untrained control (C-NT); trained control (C-T); untrained high fructose diet (HFRU-NT); trained high fructose diet (HFRU-T); untrained high fat diet (HF- NRT); trained high fat diet.

Note: data are presented as means ± SD, significant differences between groups are indicated by symbols above the bars; P <0.05, one-way ANOVA and Holm-Sidak post-test: [a] compared with C-NT; [b] different from HFRU-NT, [c] different from HF-NT.

Source: the author, 2017.

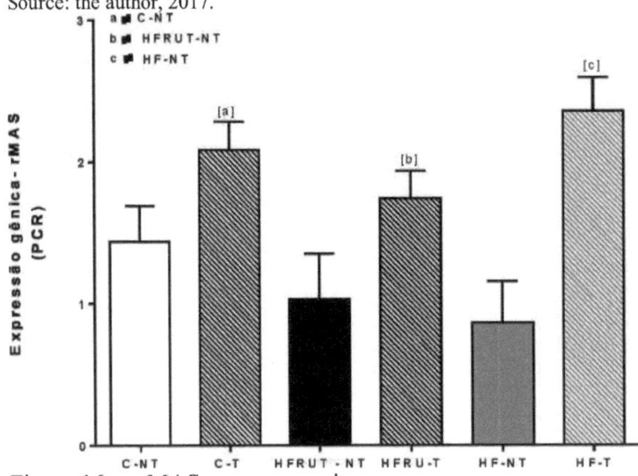

Figura 16- rMAS gene expression.

Legend: untrained control (C-NT); trained control (C-T); untrained high fructose diet (HFRU-NT); trained high fructose diet (HFRU-T); untrained high fat diet (HF- NRT); trained high fat diet.

Note: data are presented as means ± SD, significant differences between groups are indicated by symbols above the bars; P <0.05, one-way ANOVA and Holm-Sidak post-test: [a] compared with C-NT; [b] different from HFRU-NT, [c] different from HF-NT.

Source: the Author, 2017

CHAPTER 5

DISCUSSION

The present study used high-fat or high-fructose diets to induce cardiometabolic alterations in an experimental mouse model and subjected them to a HIIT protocol to investigate how left ventricular remodelling adapts with or without obesity.The results of the present study showed that the ingestion of high-fat or high-fructose diets is detrimental to insulin sensitivity, leading to glucose intolerance, blood pressure, left ventricular hypertrophy and alterations in the LV RAS axis.

On the other hand, HIIT showed beneficial cardiovascular effects. The main findings were: (I) SBP was reduced with HIIT in mice fed a high-fat diet; however, this was not observed in mice fed a high-fructose diet; (II) left ventricular hypertrophy in both groups was better after 12 weeks of HIIT sessions; (III) components of the left ventricular RAS axis showed results influenced by both diet and HIIT. For example, AT1R protein expression was predominantly influenced by an interaction between diet and HIIT, while AT2R protein expression was predominantly influenced by HIIT. ACE protein expression, on the other hand, was only influenced by diet and HIIT independently.

The literature describes that HIIT induces moderate weight loss (0.5-4kg reduction) in adults with common metabolic diseases (Tjonna et al., 2008; Madsen et al., 2015). This is in line with our results, as the HF-T group showed moderate weight loss compared to the HF-NT group, which can be explained because it is postulated that HIIT increases mitochondrial density and capacity leading to increased fat oxidation. These findings are important because these fat deposits increase the risk of cardiovascular disease (Talanian et al., 2007; Wu et al., 2016) and metabolic dysfunction (Lee et al., 2016b). In addition, the respiratory quotient (RQ) was significantly higher in the HFRU-NT group than in all other groups. The RQ varies with the substrate being predominantly oxidised, and is thus used to determine the relative contribution of glucose, lipids and proteins to energy production. Whereas energy expenditure was higher in the HF-T group, justifying the reduction in body mass in this group and the beneficial effect of HIIT on this parameter.

With regard to glucose tolerance, animals fed high-fat or high-fructose diets are well-established models used to study changes in glucose tolerance and insulin resistance, as described in previous studies by our group (Fraulob et al., 2010; Oliveira et al., 2014; Franssens et al., 2016). In the present study, 12 weeks of HIIT protocol reduced the area under the curve at the 2-hour time point after exercise overload in the HF-T and HFRU-T groups. These improvements in glucose metabolism in the HIIT groups have been associated with increased glucose uptake in skeletal muscle and improved insulin sensitivity (Schultz et al., 2015). In addition, HIIT is reported to reduce the area under the glucose curve in those with glucose intolerance (Marquis-Gravel et al., 2015; Batacan et al., 2016) or type 2 diabetes (Little et al., 2014). Controlling glucose tolerance is very important because postprandial glucose changes are strong predictors of cardiovascular disease (Terada et al., 2016), which may be due to possible inductions caused via oxidative stress.

In addition, HIIT has also been shown to improve peripheral insulin sensitivity in people with metabolic alterations; in the present study the HIIT protocol showed a positive response in insulin sensitivity in the HF-T, and HFRU- T groups. Chronic intake of a high-fat diet, as well as excessive fructose intake (Rebolledo and Actis Dato, 2005; Cavalot et al., 2011), leads to hyperglycaemia and as a consequence the installation of insulin resistance (Balakumar et *al.,* 2016). The literature describes that the improvements in insulin sensitivity with HIIT are comparable to the improvements obtained with moderate-intensity continuous resistance exercise (MCIT) (Nybo et al., 2010; Mitranun et al., 2014; Marcinko et al., 2015). Importantly, several studies have shown that HIIT can improve insulin sensitivity, independently of weight loss and adiposity, in adults who are sedentary (Tjonna et al., 2008), obese (Cocks et al., 2016) or have type 2 diabetes (Shaban et al., 2014).

Physical exercise has the potential to reduce cardiometabolic risk factors, including improvements in blood pressure (Pimenta et al., 2015; Batacan et al., 2016). In the current study, the effects of the HIIT intervention on reducing SBP were observed in the HF-T group. However, the HFRU-T group did not show the same beneficial effects after the HIIT protocol. Several mechanisms could potentially explain the hypertensive effects of fructose; there is evidence to support that hypertension induced by fructose overload leads to increased expression of AT1-R in rodents (Hollekim-Strand et al., 2014). Increased expression of AT1-R and angiotensin II has also been shown in rodents fed a high-fructose diet (Giacchetti et al., 2000). According to another theory associating diet with BP regulation (Shinozaki et al., 2004), the presence of carbohydrates in the diet would lead to increased circulating insulin concentrations, which can act centrally to stimulate the sympathetic neural activity of the heart, resulting in increased heart rate and cardiac output (Landsberg and Young, 1985).

The third factor that may explain the rise in SBP in the HFRU-NT and HFRU-T groups involving fructose intake is uric acid; increased dietary fructose intake is associated with moderate but highly significant rises in uric acid concentrations (Charriere et al., 2016). Fructose intake can stimulate endogenous uric acid synthesis (Wang et al., 2012) and elevated uric acid concentration can also impair normal endothelial function and contribute to the development of hypertension or insulin resistance (Hasenfuss, 1998). This effect is observed in high fructose diets when fructose is an isoenergetic substitute for starch or fat, in the same way as the diet in this study. Our uric acid results after fructose intake corroborate the SBP results; the HFRU-NT and HFRU-T groups had high uric acid and no reduction after HIIT, in agreement with a recent study which also found no significant reduction in uric acid after HIIT in an experimental model (Kanbay et al., 2013).

Another important finding of this study is left ventricular hypertrophy. Animals fed high-fat diets and high-fructose diets are well-established models used to study cardiac hypertrophy (Wang et al., 2015; de Araujo et al., 2016), and studies have indicated that, in the long term, 12 weeks of high fructose intake is capable of inducing ventricular hypertrophy (Bouchard-Thomassin et al., 2011). In our study, animals were fed high-fat or high-fructose diets for 20 weeks and exhibited both increased mass and significant LV hypertrophy in the HFRU-NT and HF-NT groups, but after HIIT this hypertrophy was attenuated in the trained groups. Current studies have found that the key mechanisms of cardiac hypertrophy induced by high fat and/or fructose content are hypertension (Mellor et al., 2010), insulin resistance (Frantz et al., 2013), and mitochondrial dysfunction (Raher et al., 2008; Bouchard-Thomassin et al., 2011). A high intake of dietary fat (Dong et al., 2007) and long-term dietary fructose induce systemic glucose dysregulation and cardiac hypertrophic response, including a markedly enlarged heart, and abundantly swollen cardiomyocytes.

Cardiac hypertrophy occurs in response to various stimuli such as chronic physical training (volume overload) and hypertension (pressure overload). On the other hand, the literature indicates that HIIT provides a stimulus to induce myocardial improvements and consequently confirms the beneficial impact of HIIT on vascular and cardiorespiratory fitness. This suggests that the cardiovascular benefit of HIIT outweighs its metabolic benefits. The effects of different exercise intensities on the diseased heart muscle are incompletely understood and, furthermore, the number of studies investigating cardiac structure after HIIT is small. However, there is evidence that HIIT induces physiological hypertrophy (Sverdlov et al., 2016).

With regard to protein expression, HIIT showed beneficial effects on some components of the RAS axis, such as reduced expression of ACE and AT2R in the HF-T and HFRU-T groups. However, protein expression of AT1R in the HFRU-T group was not reduced; this may be explained because hypertension induced by fructose overload leads to increased expression of AT1-R in rodents (Wang et al., 2012).

With regard to gene expression, it is worth mentioning the ACE2/Ang 1-7/rMAS axis. ACE2 gene expression was higher in the HF-T and HFRU-T groups, which was followed by a beneficial response in rMAS gene expression after HIIT in the HF-T and HFRU-T groups. These findings led us to hypothesise that the Ang-(1-7)/rMAS axis may play a role in physiological cardiac remodelling induced by diets high in fat or fructose and ameliorated after HIIT (Manrique et al., 2009; De Mello, 2017).

Recent studies have indicated that ACE2 is the main enzyme involved in the Ang-(1-7) synthesis pathway in vital organs such as the heart (Baker et al., 1990). Most Ang-(1-7) actions are mediated by the receptor, MAS, which is present in the heart and plays an important role in improving cardiac function (Fernandes et al., 2011), triggering signalling pathways that lead to vasodilation and anti-fibrotic, anti-hypertrophic and anti-arrhythmic effects (Pereira et al., 2009). In addition, a recent study showed that physical exercise reduces cardiac ACE and AngII increases cardiac ACE2 and Ang-(1-7) (Fernandes et al., 2011).

In short, our results confirm that both the high-fat diet and fructose are capable of causing glucose intolerance, hypertension and adverse cardiac remodelling, regardless of the presence or absence of obesity in the experimental model. In addition, HIIT had beneficial effects, combating insulin resistance and attenuating LV hypertrophy, although hypertension was controlled only in the animals in the high-fat group submitted to the HIIT protocol. The cardiac RAS mediates these findings and the MAS receptor seems to play a crucial role when it comes to improving cardiac structural and functional remodelling due to HIIT.

REFERENCES

Aeberli I, Gerber PA, Hochuli M, Kohler S, Haile SR, Gouni-Berthold I et al. Low to moderate sugar-sweetened beverage consumption impairs glucose and lipid metabolism and promotes inflammation in healthy young men: a randomised controlled trial. Am J Clin Nutr 2011; 94:479-485.

Alwahsh SM, Gebhardt R. Dietary fructose as a risk factor for non-alcoholic fatty liver disease (NAFLD). Arch Toxicol 2017; 91:1545-1563.

Anderson E, Shivakumar G. Effects of exercise and physical activity on anxiety. Front Psychiatry 2013; 4:27.

Anversa P, Ricci R, Olivetti G. Quantitative structural analysis of the myocardium during physiologic growth and induced cardiac hypertrophy: a review. J Am Coll Cardiol 1986; 7:1140-1149.

Bader M. Role of the local renin-angiotensin system in cardiac damage: a minireview focussing on transgenic animal models. J Mol Cell Cardiol 2002; 34:1455-1462.

Baker KM, Chernin MI, Wixson SK, Aceto JF. Renin-angiotensin system involvement in pressure-overload cardiac hypertrophy in rats. Am J Physiol 1990; 259:H324-332.

Balakumar M, Raji L, Prabhu D, Sathishkumar C, Prabu P, Mohan V et al. High- fructose diet is as detrimental as high-fat diet in the induction of insulin resistance and diabetes mediated by hepatic/pancreatic endoplasmic reticulum (ER) stress. Mol Cell Biochem 2016; 423:93-104.

BARREIROS RC, BOSSOLAN G, TRINDADE CEP. Fructose in humans: metabolic effects, clinical utilisation, and associated inherent errors. Rev. Nutr., Campinas 2005.

Barsukov AV, Glukhovskoy DV, Zobnina MP, Mirokhina MA, Dydyshko VT, Vasiliev VN et al. Left Ventricular Hypertrophy as a Marker of Adverse Cardiovascular Risk in Persons of Different Age Groups. Advances in Gerontology 2015; 5:99-106.

Bassuk SS, Manson JE. Epidemiological evidence for the role of physical activity in reducing risk of type 2 diabetes and cardiovascular disease. J Appl Physiol (1985) 2005; 99:1193-1204.

Batacan RB, Jr, Duncan MJ, Dalbo VJ, Connolly KJ, Fenning AS. Light- intensity and high-intensity interval training improve cardiometabolic health in rats. Appl Physiol Nutr Metab 2016; 41:945-952.

Bessesen DH. Update on obesity. J Clin Endocrinol Metab 2008; 93:2027-2034. Bidwell AJ, Fairchild TJ, Redmond J, Wang L, Keslacy S, Kanaley JA. Physical activity offsets the negative effects of a high-fructose diet. Med Sci Sports Exerc 2014; 46:2091-2098.

Bizeau ME, Pagliassotti MJ. Hepatic adaptations to sucrose and fructose. Metabolism 2005; 54:1189-1201.

Borges JP, Masson GS, Tibirica E, Lessa MA. Aerobic interval exercise training induces greater reduction in cardiac workload in the recovery period in rats. Arq Bras Cardiol 2014; 102:47-53.
Bouchard-Thomassin AA, Lachance D, Drolet MC, Couet J, Arsenault M. A high-fructose diet worsens eccentric left ventricular hypertrophy in experimental volume overload. Am J Physiol Heart Circ Physiol 2011; 300:H125-134.

Boudet G, Walther G, Courteix D, Obert P, Lesourd B, Pereira B et al. Paradoxical dissociation between heart rate and heart rate variability following different modalities of exercise in individuals with metabolic syndrome: The RESOLVE study. Eur J Prev Cardiol 2016; 10.1177/2047487316679523.
Brouwer IA, Wanders AJ, Katan MB. Effect of animal and industrial trans fatty acids on HDL and LDL cholesterol levels in humans--a quantitative review. PLoS One 2010; 5:e9434.
Burgomaster KA, Howarth KR, Phillips SM, Rakobowchuk M, Macdonald MJ, McGee SL et al. Similar metabolic adaptations during exercise after low volume sprint interval and traditional endurance training in humans. J Physiol 2008; 586:151-160.
Burgomaster KA, Hughes SC, Heigenhauser GJ, Bradwell SN, Gibala MJ. Six sessions of sprint interval training increases muscle oxidative potential and cycle endurance capacity in humans. J Appl Physiol (1985) 2005; 98:1985- 1990.
Cassidy S, Thoma C, Hallsworth K, Parikh J, Hollingsworth KG, Taylor R et al. High intensity intermittent exercise improves cardiac structure and function and reduces liver fat in patients with type 2 diabetes: a randomised controlled trial. Diabetologia 2016; 59:56-66.
Cassidy S, Thoma C, Houghton D, Trenell MI. High-intensity interval training: a review of its impact on glucose control and cardiometabolic health. Diabetologia 2017; 60:7-23.
Cavalot F, Pagliarino A, Valle M, Di Martino L, Bonomo K, Massucco P et al. Postprandial blood glucose predicts cardiovascular events and all-cause mortality in type 2 diabetes in a 14-year follow-up: lessons from the San Luigi Gonzaga Diabetes Study. Diabetes Care 2011; 34:2237-2243.
Charriere N, Loonam C, Montani JP, Dulloo AG, Grasser EK. Cardiovascular responses to sugary drinks in humans: galactose presents milder cardiac effects than glucose or fructose. Eur J Nutr 2016; 10.1007/s00394-016-1250-9. Cheeseman CI. GLUT2 is the transporter for fructose across the rat intestinal basolateral membrane. Gastroenterology 1993; 105:1050-1056.
Chinnaiyan KM, Alexander D, McCullough PA. Role of angiotensin II in the evolution of diastolic heart failure. J Clin Hypertens (Greenwich) 2005; 7:740- 747.
Cocks M, Shaw CS, Shepherd SO, Fisher JP, Ranasinghe A, Barker TA et al. Sprint interval and moderate-intensity continuous training have equal benefits on aerobic capacity, insulin sensitivity, muscle capillarisation and endothelial eNOS/NAD(P)Hoxidase protein ratio in obese men. J Physiol 2016; 594:2307- 2321.
Colville CA, Seatter MJ, Jess TJ, Gould GW, Thomas HM. Kinetic analysis of the liver-type (GLUT2) and brain-type (GLUT3) glucose transporters in Xenopus oocytes: substrate specificities and effects of transport inhibitors. Biochem J 1993; 290 (Pt 3):701-706.
de Araujo GG, Papoti M, Dos Reis IG, de Mello MA, Gobatto CA. Short and Long Term Effects of High-Intensity Interval Training on Hormones, Metabolites, Antioxidant System, Glycogen Concentration, and Aerobic Performance Adaptations in Rats. Front Physiol 2016; 7:505.
De Mello WC. Local Renin Angiotensin Aldosterone Systems and Cardiovascular Diseases. Med Clin North Am 2017; 101:117-127.
Domenighetti AA, Wang Q, Egger M, Richards SM, Pedrazzini T, Delbridge LM. Angiotensin II-mediated phenotypic cardiomyocyte remodelling leads to age- dependent cardiac dysfunction and failure. Hypertension 2005; 46:426-432. Dong F, Li Q, Sreejayan N, Nunn JM, Ren J. Metallothionein prevents high-fat diet induced cardiac contractile

dysfunction: role of peroxisome proliferator activated receptor gamma coactivator 1alpha and mitochondrial biogenesis. Diabetes 2007; 56:2201-2212.
Donoghue M, Hsieh F, Baronas E, Godbout K, Gosselin M, Stagliano N et al. A novel angiotensin-converting enzyme-related carboxypeptidase (ACE2) converts angiotensin I to angiotensin 1-9. Circ Res 2000; 87:E1-9.
Douard V, Ferraris RP. Regulation of the fructose transporter GLUT5 in health and disease. Am J Physiol Endocrinol Metab 2008; 295:E227-237.
Erridge C, Attina T, Spickett CM, Webb DJ. A high-fat meal induces low-grade endotoxemia: evidence of a novel mechanism of postprandial inflammation. Am J Clin Nutr 2007; 86:1286-1292.
Fernandes T, Hashimoto NY, Magalhaes FC, Fernandes FB, Casarini DE, Carmona AK et al. Aerobic exercise training-induced left ventricular hypertrophy involves regulatory MicroRNAs, decreased angiotensin-converting enzyme- angiotensin ii, and synergistic regulation of angiotensin-converting enzyme 2- angiotensin (1-7). Hypertension 2011; 58:182-189.
Fex A, Leduc-Gaudet JP, Filion ME, Karelis AD, Aubertin-Leheudre M. Effect of Elliptical High Intensity Interval Training on Metabolic Risk Factor in Pre- and Type 2 Diabetes Patients: A Pilot Study. J Phys Act Health 2015; 12:942-946. Fisher G, Brown AW, Bohan Brown MM, Alcorn A, Noles C, Winwood L et al. High Intensity Interval- vs Moderate Intensity- Training for Improving Cardiometabolic Health in Overweight or Obese Males: A Randomised Controlled Trial. PLoS One 2015; 10:e0138853.
Flores-Munoz M, Godinho BM, Almalik A, Nicklin SA. Adenoviral delivery of angiotensin-(1-7) or angiotensin-(1-9) inhibits cardiomyocyte hypertrophy via the mas or angiotensin type 2 receptor. PLoS One 2012; 7:e45564.
Franchini KG. Cardiac hypertrophy: molecular mechanisms. Brazilian journal of hypertension 2001; 8:125-141.
Franssens BT, Westerink J, van der Graaf Y, Nathoe HM, Visseren FL, group Ss. Metabolic consequences of adipose tissue dysfunction and not adiposity per se increase the risk of cardiovascular events and mortality in patients with type 2 diabetes. Int J Cardiol 2016; 222:72-77.
Frantz ED, Crespo-Mascarenhas C, Barreto-Vianna AR, Aguila MB, Mandarim- de-Lacerda CA. Renin-angiotensin system blockers protect pancreatic islets against diet-induced obesity and insulin resistance in mice. PLoS One 2013; 8:e67192.
Fraulob JC, Ogg-Diamantino R, Fernandes-Santos C, Aguila MB, Mandarim-de- Lacerda CA. A Mouse Model of Metabolic Syndrome: Insulin Resistance, Fatty Liver and Non-Alcoholic Fatty Pancreas Disease (NAFPD) in C57BL/6 Mice Fed a High Fat Diet. J Clin Biochem Nutr 2010; 46:212-223.
Frey N, Katus HA, Olson EN, Hill JA. Hypertrophy of the heart: a new therapeutic target? Circulation 2004; 109:1580-1589.
Funaki M. Saturated fatty acids and insulin resistance. J Med Invest 2009; 56:88-92.
Garber CE, Blissmer B, Deschenes MR, Franklin BA, Lamonte MJ, Lee IM et al. American College of Sports Medicine position stand. Quantity and quality of exercise for developing and maintaining cardiorespiratory, musculoskeletal, and neuromotor fitness in apparently healthy adults: guidance for prescribing exercise. Med Sci Sports Exerc 2011; 43:1334-1359.
Giacchetti G, Sechi LA, Griffin CA, Don BR, Mantero F, Schambelan M. The tissue renin-angiotensin system in rats with fructose-induced hypertension: overexpression of type 1 angiotensin II receptor in adipose tissue. J Hypertens 2000; 18:695-702.
Gibala MJ, Little JP, Macdonald MJ, Hawley JA. Physiological adaptations to low-volume, high-intensity interval training in health and disease. J Physiol 2012; 590:1077-1084.
Giles TD. Renin-angiotensin system modulation for treatment and prevention of cardiovascular diseases: towards an optimal therapeutic strategy. Rev Cardiovasc Med 2007; 8 Suppl 2:S14-21.

Hall JE, Guyton AC, Mizelle HL. Role of the renin-angiotensin system in control of sodium excretion and arterial pressure. Acta Physiol Scand Suppl 1990; 591:48-62.
Hallfrisch J. Metabolic effects of dietary fructose. PHASEB J 1990; 4:2652-2660. Hasenfuss G. Animal models of human cardiovascular disease, heart failure and hypertrophy. Cardiovasc Res 1998; 39:60-76.
Havel PJ. Dietary fructose: implications for dysregulation of energy homeostasis and lipid/carbohydrate metabolism. Nutr Rev 2005; 63:133-157.
Hein L, Meinel L, Pratt RE, Dzau VJ, Kobilka BK. Intracellular trafficking of angiotensin II and its AT1 and AT2 receptors: evidence for selective sorting of receptor and ligand. Mol Endocrinol 1997; 11:1266-1277.
Hollekim-Strand SM, Bjorgaas MR, Albrektsen G, Tjonna AE, Wisloff U, Ingul CB. High-intensity interval exercise effectively improves cardiac function in patients with type 2 diabetes mellitus and diastolic dysfunction: a randomized controlled trial. J Am Coll Cardiol 2014; 64:1758-1760.
Hu FB, van Dam RM, Liu S. Diet and risk of Type II diabetes: the role of types of fat and carbohydrate. Diabetologia 2001; 44:805-817.
Huang JP, Cheng ML, Wang CH, Shiao MS, Chen JK, Hung LM. High-fructose and high-fat feeding correspondingly lead to the development of lysoPC- associated apoptotic cardiomyopathy and adrenergic signalling-related cardiac hypertrophy. Int J Cardiol 2016; 215:65-76.
Johnson RJ, Segal MS, Sautin Y, Nakagawa T, Feig DI, Kang DH et al.
Potential role of sugar (fructose) in the epidemic of hypertension, obesity and the metabolic syndrome, diabetes, kidney disease, and cardiovascular disease. Am J Clin Nutr 2007; 86:899-906.
Kalla M, Herring N, Paterson DJ. Cardiac sympatho-vagal balance and ventricular arrhythmia. Auton Neurosci 2016; 199:29-37.
Kanbay M, Segal M, Afsar B, Kang DH, Rodriguez-Iturbe B, Johnson RJ. The role of uric acid in the pathogenesis of human cardiovascular disease. Heart 2013; 99:759-766.
Kanuri G, Spruss A, Wagnerberger S, Bischoff SC, Bergheim I. Role of tumour necrosis factor alpha (TNFalpha) in the onset of fructose-induced nonalcoholic fatty liver disease in mice. J Nutr Biochem 2011; 22:527-534.
Katz A, Nambi SS, Mather K, Baron AD, Follmann DA, Sullivan G et al.
Quantitative insulin sensitivity check index: a simple, accurate method for assessing insulin sensitivity in humans. J Clin Endocrinol Metab 2000; 85:2402- 2410.
Katz AM, Rolett EL. Heart failure: when form fails to follow function. Eur Heart J 2016; 37:449-454.
Kemi OJ, Haram PM, Loennechen JP, Osnes JB, Skomedal T, Wisloff U et al. Moderate vs. high exercise intensity: differential effects on aerobic fitness, cardiomyocyte contractility, and endothelial function. Cardiovasc Res 2005; 67:161-172.
Keys A. Atherosclerosis: a problem in newer public health. J Mt Sinai Hosp N Y 1953; 20:118-139.
Keys A. Coronary heart disease, serum cholesterol, and the diet. Acta Med Scand 1980; 207:153-160.
Ko M, Kim MT, Nam JJ. Assessing risk factors of coronary heart disease and its risk prediction among Korean adults: the 2001 Korea National Health and Nutrition Examination Survey. Int J Cardiol 2006; 110:184-190.
Lakka TA, Laaksonen DE, Lakka HM, Mannikko N, Niskanen LK, Rauramaa R et al. Sedentary lifestyle, poor cardiorespiratory fitness, and the metabolic syndrome. Med Sci Sports Exerc 2003; 35:1279-1286.
Landsberg L, Young JB. Insulin-mediated glucose metabolism in the relationship between dietary intake and sympathetic nervous system activity. Int J Obes 1985; 9 Suppl 2:63-68.
Leamy AK, Egnatchik RA, Young JD. Molecular mechanisms and the role of saturated fatty

acids in the progression of non-alcoholic fatty liver disease. Prog Lipid Res 2013; 52:165-174.
Lee J, Kim Y, Jeon JY. Association between physical activity and the prevalence of metabolic syndrome: from the Korean National Health and Nutrition Examination Survey, 1999-2012. Springerplus 2016a; 5:1870.
Lee JJ, Pedley A, Hoffmann U, Massaro JM, Fox CS. Association of Changes in Abdominal Fat Quantity and Quality With Incident Cardiovascular Disease Risk Factors. J Am Coll Cardiol 2016b; 68:1509-1521.
Levy D, Garrison RJ, Savage DD, Kannel WB, Castelli WP. Prognostic implications of echocardiographically determined left ventricular mass in the Framingham Heart Study. N Engl J Med 1990; 322:1561-1566.
Liang Y, Huang B, Song E, Bai B, Wang Y. Constitutive activation of AMPK alpha1 in vascular endothelium promotes high-fat diet-induced fatty liver injury: role of COX-2 induction. Br J Pharmacol 2014; 171:498-508.
Little JP, Jung ME, Wright AE, Wright W, Manders RJ. Effects of high-intensity interval exercise versus continuous moderate-intensity exercise on postprandial glycemic control assessed by continuous glucose monitoring in obese adults. Appl Physiol Nutr Metab 2014; 39:835-841.
Macotela Y, Boucher J, Tran TT, Kahn CR. Sex and depot differences in adipocyte insulin sensitivity and glucose metabolism. Diabetes 2009; 58:803- 812.
Madsen SM, Thorup AC, Overgaard K, Jeppesen PB. High Intensity Interval Training Improves Glycaemic Control and Pancreatic beta Cell Function of Type 2 Diabetes Patients. PLoS One 2015; 10:e0133286.
Mangold S, Kramer U, Franzen E, Erz G, Bretschneider C, Seeger A et al. Detection of cardiovascular disease in elite athletes using cardiac magnetic resonance imaging. Rofo 2013; 185:1167-1174.
Manrique C, Lastra G, Gardner M, Sowers JR. The renin angiotensin aldosterone system in hypertension: roles of insulin resistance and oxidative stress. Med Clin North Am 2009; 93:569-582.
Mansur AP, Favarato D. Mortality from cardiovascular diseases in Brazil and the São Paulo metropolitan region: 2011 update. rquivos brasieliros de cardiologia 2012; 99:755/761.
Marcinko K, Sikkema SR, Samaan MC, Kemp BE, Fullerton MD, Steinberg GR. High-intensity interval training improves liver and adipose tissue insulin sensitivity. Mol Metab 2015; 4:903-915.
Marquis-Gravel G, Hayami D, Juneau M, Nigam A, Guilbeault V, Latour E et al. Intensive lifestyle intervention including high-intensity interval training program improves insulin resistance and fasting plasma glucose in obese patients. Prev Med Rep 2015; 2:314-318.
Matos-Souza JR, Franchini KG, Nadruz Junior W. Left ventricular hypertrophy: pathway to heart failure. Brazilian journal of hypertension 2008; 15:71-74.
Mellor K, Ritchie RH, Meredith G, Woodman OL, Morris MJ, Delbridge LM. High-fructose diet elevates myocardial superoxide generation in mice in the absence of cardiac hypertrophy. Nutrition 2010; 26:842-848.
Mihl C, Dassen WR, Kuipers H. Cardiac remodelling: concentric versus eccentric hypertrophy in strength and endurance athletes. Neth Heart J 2008; 16:129-133.
Mill JG, Vassallo DV. Cardiac hypertrophy. Brazilian journal of hypertension 2001; 8:63-75.
Miller M, Stone NJ, Ballantyne C, Bittner V, Criqui MH, Ginsberg HN et al. Triglycerides and cardiovascular disease: a scientific statement from the American Heart Association. Circulation 2011; 123:2292-2333.
Mirza MS. Obesity, Visceral Fat, and NAFLD: Querying the Role of Adipokines in the Progression of Nonalcoholic Fatty Liver Disease. ISRN Gastroenterol 2011; 2011:592404.
Mitranun W, Deerochanawong C, Tanaka H, Suksom D. Continuous vs interval training on glycemic control and macro- and microvascular reactivity in type 2 diabetic patients. Scand J

Med Sci Sports 2014; 24:e69-76.
Mochcovitch MD, Deslandes AC, Freire RC, Garcia RF, Nardi AE. The effects of regular physical activity on anxiety symptoms in healthy older adults: a systematic review. revista brasileira de psiquiatria 2016; 38:45-51.
Molmen-Hansen HE, Stolen T, Tjonna AE, Aamot IL, Ekeberg IS, Tyldum GA et al. Aerobic interval training reduces blood pressure and improves myocardial function in hypertensive patients. Eur J Prev Cardiol 2012; 19:151-160.
Mozaffarian D, Appel LJ, Van Horn L. Components of a cardioprotective diet: new insights. Circulation 2011; 123:2870-2891.
MS. Hypertension. Ministry of Health: Primary Care Notebook 2006; I:7-51.
Nadruz W. Myocardial remodelling in hypertension. J Hum Hypertens 2015; 29:1-6.
Nakagawa T, Hu H, Zharikov S, Tuttle KR, Short RA, Glushakova O et al. A causal role for uric acid in fructose-induced metabolic syndrome. Am J Physiol Renal Physiol 2006; 290:F625-631.
Nicolo A, Girardi M. The physiology of interval training: a new target to HIIT. J Physiol 2016; 594:7169-7170.
Nybo L, Sundstrup E, Jakobsen MD, Mohr M, Hornstrup T, Simonsen L et al. High-intensity training versus traditional exercise interventions for promoting health. Med Sci Sports Exerc 2010; 42:1951-1958.
Oliveira LS, Santos DA, Barbosa-da-Silva S, Mandarim-de-Lacerda CA, Aguila MB. The inflammatory profile and liver damage of a sucrose-rich diet in mice. J Nutr Biochem 2014; 25:193-200.
Passos-Silva DG, Brandan E, Santos RA. Angiotensins as therapeutic targets beyond heart disease. Trends Pharmacol Sci 2015; 36:310-320.
Pereira MG, Ferreira JC, Bueno CR, Jr, Mattos KC, Rosa KT, Irigoyen MC et al. Exercise training reduces cardiac angiotensin II levels and prevents cardiac dysfunction in a genetic model of sympathetic hyperactivity-induced heart failure in mice. Eur J Appl Physiol 2009; 105:843-850.
Pierard M, Conotte S, Tassin A, Boutry S, Uzureau P, Boudjeltia KZ et al. Interactions of exercise training and high-fat diet on adiponectin forms and muscle receptors in mice. Nutr Metab (Lond) 2016; 13:75.
Pimenta M, Bringhenti I, Souza-Mello V, Dos Santos Mendes IK, Aguila MB, Mandarim-de-Lacerda CA. High-intensity interval training beneficial effects on body mass, blood pressure, and oxidative stress in diet-induced obesity in ovariectomised mice. Life Sci 2015; 139:75-82.
Pinto JR VC, Daher CV, Sallum FS, Jatene MB, Croti UA. The situation of congenital heart surgeries in Brazil: Pediatric Cardiovascular Surgery Department (PCVSD) of the Brazilian Society of Cardiovascular Surgery (BSCVS).
Brazilian journal of cardiovascular surgery 2004; 19:III-VI.
Radovits T, Olah A, Lux A, Nemeth BT, Hidi L, Birtalan E et al. Rat model of exercise-induced cardiac hypertrophy: haemodynamic characterization using left ventricular pressure-volume analysis. Am J Physiol Heart Circ Physiol 2013; 305:H124-134.
Rafiq N, Bai C, Fang Y, Srishord M, McCullough A, Gramlich T et al. Long-term follow-up of patients with nonalcoholic fatty liver. Clin Gastroenterol Hepatol 2009; 7:234-238.
Raher MJ, Thibault HB, Buys ES, Kuruppu D, Shimizu N, Brownell AL et al. A short duration of high-fat diet induces insulin resistance and predisposes to adverse left ventricular remodelling after pressure overload. Am J Physiol Heart Circ Physiol 2008; 295:H2495-2502.
Rebolledo OR, Actis Dato SM. Postprandial hyperglycemia and hyperlipidemia- generated glycoxidative stress: its contribution to the pathogenesis of diabetes complications. Eur Rev Med Pharmacol Sci 2005; 9:191-208.
Reeves PG, Nielsen FH, Fahey GC, Jr. AIN-93 purified diets for laboratory rodents: final report of the American Institute of Nutrition ad hoc writing committee on the reformulation of the AIN-76A rodent diet. J Nutr 1993; 123:1939-1951.

Rice GI, Thomas DA, Grant PJ, Turner AJ, Hooper NM. Evaluation of angiotensin-converting enzyme (ACE), its homologue ACE2 and neprilysin in angiotensin peptide metabolism. Biochem J 2004; 383:45-51.
Rippe JM, Angelopoulos TJ. Added sugars and risk factors for obesity, diabetes and heart disease. Int J Obes (Lond) 2016a; 40 Suppl 1:S22-27.
Rippe JM, Angelopoulos TJ. Sugars, obesity, and cardiovascular disease: results from recent randomised control trials. Eur J Nutr 2016b; 55:45-53.
Sacks FM, Lichtenstein AH, Wu JHY, Appel LJ, Creager MA, Kris-Etherton PM et al. Dietary Fats and Cardiovascular Disease: A Presidential Advisory From the American Heart Association. Circulation 2017; 136:e1-e23.
Samir N, Mahmud S, Khuwaja AK. Prevalence of physical inactivity and barriers to physical activity among obese attendants at a community health-care centre in Karachi, Pakistan. BMC Res Notes 2011; 4:174.
Santos RA, Simoes e Silva AC, Maric C, Silva DM, Machado RP, de Buhr I et al. Angiotensin-(1-7) is an endogenous ligand for the G protein-coupled receptor Mas. Proc Natl Acad Sci U S A 2003; 100:8258-8263.
SBC. VI Brazilian Hypertension Guidelines. Arqueivos brasileiros de cardiologia 2010; 95:1-51.
SBC. 2015. Cardiometer. In: Cardiometer: death from cardiovascular disease in Brazil. http://www.cardiometro.com.br/anteriores.asp: Brazilian Society of Cardiology.
SBC. VII Brazilian hypertension guideline. Arquivos brasileiros de cardiologia 2016; 103:1-82.
Scharf M, Brem MH, Wilhelm M, Schoepf UJ, Uder M, Lell MM. Atrial and ventricular functional and structural adaptations of the heart in elite triathletes assessed with cardiac MR imaging. Radiology 2010; 257:71-79.
Schultz A, Barbosa-da-Silva S, Aguila MB, Mandarim-de-Lacerda CA. Differences and similarities in hepatic lipogenesis, gluconeogenesis and oxidative imbalance in mice fed diets rich in fructose or sucrose. Food Funct 2015; 6:1684-1691.
Sciarretta S, Paneni F, Ciavarella GM, De Biase L, Palano F, Baldini R et al. Evaluation of systolic properties in hypertensive patients with different degrees of diastolic dysfunction and normal ejection fraction. Am J Hypertens 2009; 22:437-443.
Serneri GG, Boddi M, Cecioni I, Vanni S, Coppo M, Papa ML et al. Cardiac angiotensin II formation in the clinical course of heart failure and its relationship with left ventricular function. Circ Res 2001; 88:961-968.
Shaban N, Kenno KA, Milne KJ. The effects of a 2 week modified high intensity interval training programme on the homeostatic model of insulin resistance (HOMA-IR) in adults with type 2 diabetes. J Sports Med Phys Fitness 2014; 54:203-209.
Shinozaki K, Ayajiki K, Nishio Y, Sugaya T, Kashiwagi A, Okamura T. Evidence for a causal role of the renin-angiotensin system in vascular dysfunction associated with insulin resistance. Hypertension 2004; 43:255-262.
SIGN. Risk estimation and the prevention of cardiovascular disease: A national clinical guideline. Scottish Intercollegiate Guidelines Network 2007; I:1-71.
Simoes e Silva AC, Silveira KD, Ferreira AJ, Teixeira MM. ACE2, angiotensin- (1-7) and Mas receptor axis in inflammation and fibrosis. Br J Pharmacol 2013; 169:477-492.
Steckelings UM, Rompe F, Kaschina E, Unger T. The evolving story of the RAAS in hypertension, diabetes and CV disease: moving from macrovascular to microvascular targets. Fundam Clin Pharmacol 2009; 23:693-703.
Stolen TO, Hoydal MA, Kemi OJ, Catalucci D, Ceci M, Aasum E et al. Interval training normalises cardiomyocyte function, diastolic Ca2+ control, and SR Ca2+ release synchronicity in a mouse model of diabetic cardiomyopathy. Circ Res 2009; 105:527-536.
Sverdlov AL, Elezaby A, Qin F, Behring JB, Luptak I, Calamaras TD et al. Mitochondrial Reactive Oxygen Species Mediate Cardiac Structural, Functional, and Mitochondrial

Consequences of Diet-Induced Metabolic Heart Disease. J Am Heart Assoc 2016; 5.
Talanian JL, Galloway SD, Heigenhauser GJ, Bonen A, Spriet LL. Two weeks of high-intensity aerobic interval training increases the capacity for fat oxidation during exercise in women. J Appl Physiol (1985) 2007; 102:1439-1447.
Tappy L, Le KA. Metabolic effects of fructose and the worldwide increase in obesity. Physiol Rev 2010; 90:23-46.
Terada T, Wilson BJ, Myette-Comicronte E, Kuzik N, Bell GJ, McCargar LJ et al. Targeting specific interstitial glycemic parameters with high-intensity interval exercise and fasted-state exercise in type 2 diabetes. Metabolism 2016; 65:599- 608.
Tjonna AE, Lee SJ, Rognmo O, Stolen TO, Bye A, Haram PM et al. Aerobic interval training versus continuous moderate exercise as a treatment for the metabolic syndrome: a pilot study. Circulation 2008; 118:346-354.
Topping DL, Mayes PA. The immediate effects of insulin and fructose on the metabolism of the perfused liver. Changes in lipoprotein secretion, fatty acid oxidation and esterification, lipogenesis and carbohydrate metabolism. Biochem J 1972; 126:295-311.
Touyz RM, Berry C. Recent advances in angiotensin II signalling. Braz J Med Biol Res 2002; 35:1001-1015.
Unger RH. Minireview: weapons of lean body mass destruction: the role of ectopic lipids in the metabolic syndrome. Endocrinology 2003; 144:5159-5165. van Kats JP, Danser AH, van Meegen JR, Sassen LM, Verdouw PD, Schalekamp MA. Angiotensin production by the heart: a quantitative study in pigs with the use of radiolabelled angiotensin infusions. Circulation 1998; 98:73- 81.
Vehi C, Falces C, Sarlat MA, Gonzalo A, Andrea R, Sitges M. Nordic walking for cardiovascular prevention in patients with ischaemic heart disease or metabolic syndrome. Med Clin (Barc) 2016; 147:537-539.
Vickers C, Hales P, Kaushik V, Dick L, Gavin J, Tang J et al. Hydrolysis of biological peptides by human angiotensin-converting enzyme-related carboxypeptidase. J Biol Chem 2002; 277:14838-14843.
Wang DD, Sievenpiper JL, de Souza RJ, Chiavaroli L, Ha V, Cozma AI et al. The effects of fructose intake on serum uric acid vary among controlled dietary trials. J Nutr 2012; 142:916-923.
Wang Z, Li L, Zhao H, Peng S, Zuo Z. Chronic high fat diet induces cardiac hypertrophy and fibrosis in mice. Metabolism 2015; 64:917-925.
Whalley GA, Doughty RN, Gamble GD, Oxenham HC, Walsh HJ, Reid IR et al. Association of fat-free mass and training status with left ventricular size and mass in endurance-trained athletes. J Am Coll Cardiol 2004; 44:892-896.
Willett WC. Dietary fats and coronary heart disease. J Intern Med 2012; 272:13- 24.
Wisloff U, Loennechen JP, Currie S, Smith GL, Ellingsen O. Aerobic exercise reduces cardiomyocyte hypertrophy and increases contractility, Ca2+ sensitivity and SERCA-2 in rat after myocardial infarction. Cardiovasc Res 2002; 54:162- 174.
Wisloff U, Stoylen A, Loennechen JP, Bruvold M, Rognmo O, Haram PM et al. Superior cardiovascular effect of aerobic interval training versus moderate continuous training in heart failure patients: a randomised study. Circulation 2007; 115:3086-3094.
Wolf A, Bray GA, Popkin BM. A short history of beverages and how our body treats them. Obes Rev 2008; 9:151-164.
Wu FZ, Wu CC, Kuo PL, Wu MT. Differential impacts of cardiac and abdominal ectopic fat deposits on cardiometabolic risk stratification. BMC Cardiovasc Disord 2016; 16:20.
Xu J, Carretero OA, Liao TD, Peng H, Shesely EG, Liu TS et al. Local angiotensin II aggravates cardiac remodelling in hypertension. Am J Physiol Heart Circ Physiol 2010; 299:H1328-1338.
Yin FC, Spurgeon HA, Rakusan K, Weisfeldt ML, Lakatta EG. Use of tibial length to quantify cardiac hypertrophy: application in the aging rat. Am J Physiol 1982; 243:H941-947.

Yudkin J. Dietary Carbohydrate and Ischemic Heart Disease. Am Heart J 1963; 66:835-836.
Zile MR, Gottdiener JS, Hetzel SJ, McMurray JJ, Komajda M, McKelvie R et al. Prevalence and significance of alterations in cardiac structure and function in patients with heart failure and a preserved ejection fraction. Circulation 2011; 124:2491-2501.

Annex A - Certificate from the Ethics Committee for the care and use of experimental animals

COMISSÃO DE ÉTICA PARA O CUIDADO E USO DE ANIMAIS EXPERIMENTAIS (CEUA)

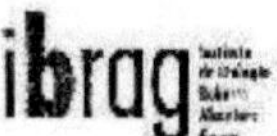

CERTIFICADO

Certificamos que o Protocolo nº **CEUA/013/2016** sobre "**Efeito do treinamento de alta intensidade no estresse oxidativo e alterações metabólicas e estruturais cardiovasculares em modelo de dieta rica em frutose e rica em lipídio**" sob a responsabilidade de **Sandra Barbosa da Silva**, está de acordo com os Princípios Éticos na Experimentação Animal adotados pelo Conselho Nacional de Controle de Experimentação Animal (CONCEA), tendo sido aprovado pela Comissão de Ética Para o Cuidado e Uso de Animais Experimentais do Instituto de Biologia Roberto Alcântara Gomes da UERJ (CEUA), em **29/03/2016**. Este certificado expira em **29/03/2020**.

Rio de Janeiro, 29 de Março de 2016.

Prof. Dr. Alex C. Manhães
Coordenador
CEUA/IBRAG/UERJ

Profa. Dra. Patricia C. Lisboa
Vice-Coordenadora
CEUA/IBRAG/UERJ

http://www.biologiauerj.com.br/comite-de-etica
ceua.ibrag@yahoo.com.br

Annex B - Article submission

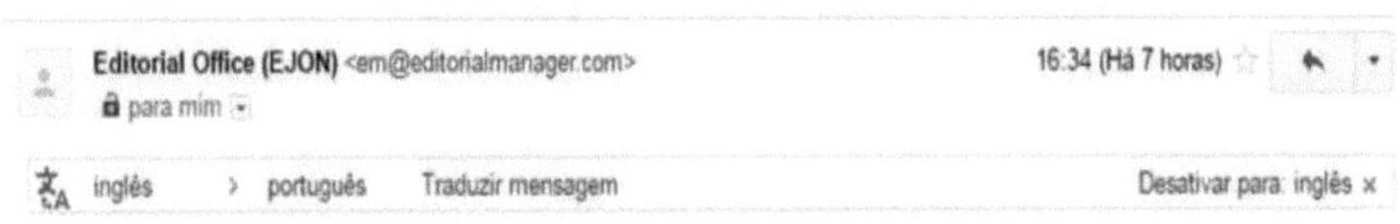

EJON-D-17-00051 - Submission Notification to co-author Entrada x

Editorial Office (EJON) <em@editorialmanager.com> 16:34 (Há 7 horas)

para mim

inglês > português Traduzir mensagem Desativar para: inglês x

Re: "High-intensity interval training has beneficial effects on cardiac remodeling through local renin-angiotensin system modulation in mice fed high-fat or high-fructose diets"
Full author list: Guilherme de Oliveira Sá, MSC; Vivian dos Santos Neves; Shyrlei R.de Oliveira Fraga; Vanessa Souza-Mello, PhD; Sandra Barbosa-da-Silva, Ph.D

Dear Mr Sá,

We have received the submission entitled: "High-intensity interval training has beneficial effects on cardiac remodeling through local renin-angiotensin system modulation in mice fed high-fat or high-fructose diets" for possible publication in European Journal of Nutrition, and you are listed as one of the co-authors.

The manuscript has been submitted to the journal by Dr. Dr Sandra Barbosa-da-Silva who will be able to track the status of the paper through his/her login.

If you have any objections, please contact the editorial office as soon as possible. If we do not hear back from you, we will assume you agree with your co-authorship.

Thank you very much.

With kind regards,

Springer Journals Editorial Office
European Journal of Nutrition

Printed by Books on Demand GmbH, Norderstedt / Germany